THE
ART OF
WELDING

THE
ART OF
WELDING

WILLIAM L. GALVERY, JR.
FORMERLY PROFESSOR OF WELDING TECHNOLOGY,
ORANGE COAST COLLEGE, CALIFORNIA

FEATURING
RYAN FRIEDLINGHAUS
FOUNDER AND CEO, WEST COAST CUSTOMS

INDUSTRIAL PRESS

A full catalog record for this book is available from the Library of Congress.

Standard Edition: ISBN 978-0-8311-3475-4

Praxair Edition: ISBN 978-0-8311-3488-4

STATEMENT OF NON-LIABILITY

The process of welding is inherently hazardous. Welding is often used to assemble buildings, structures, vehicles, and devices whose failure can lead to embarrassment, property damage, injury, or death. The authors have carefully checked the information in this book, have had others with professional welding experience and credentials review it as well and believe that it is correct and in agreement with welding industry practices and standards. However, the authors cannot provide for every possibility, contingency, and the actions of others. If you decide to apply the information, procedures, and advice in this book, the authors are not responsible for your actions. If you decide to perform welding, use common sense, have a competent person check your designs, the welding processes, and the completed welds themselves. Seek the help of people with professional welding inspection credentials to check critical welds until you gain experience. Do not attempt life-critical welds without competent, on-the-spot assistance.

Industrial Press, Inc.
32 Haviland Street
South Norwalk, Connecticut 06854

Sponsoring Editor: John Carleo

Developmental Editor: Francis J. Donegan

Interior Text Designer: Janet Romano

Cover Designer: Paul White

Front cover photograph provided by Praxair, Inc.
Interior and back cover photographs provided by West Coast Customs, unless otherwise cited.

10 9 8 7 6 5 4 3 2 1

TABLE OF CONTENTS

Detailed Table of Contents

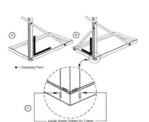

BIOGRAPHIES

Ryan Friedlinghaus is CEO and owner of West Coast Customs (WCC), a world-renowned company that builds creative and innovative custom vehicles that stretch the imagination. In 1993 Ryan's grandfather, Edward Cifranic, loaned him $5,000, and WCC was born as a modest car customization shop in Orange County, California, which then quickly grew into a larger enterprise. In 2004 WCC gained celebrity status when the MTV show *Pimp my Ride* aired, and WCC became the exclusive builder of all the vehicles for the original episodes for seasons 1- 4.

Ryan then turned his attention to expanding WCC globally, franchising shops in diverse locations such as Mexico, Germany, Malaysia, Russia, and Japan.

Following this global expansion, Ryan launched a new TV show that demonstrated that WCC was in fact a creator of automotive art, not simply a custom car shop. The result was *Street Customs and Street Customs Berlin,* which rose to the top of the ratings on the TLC network. He then took his TV career to the next level as the producer and creator of *Inside West Coast Customs,* initially airing on the Discovery channel and then on the new network, Velocity. In 2013 *Inside West Coast Customs* moved to the Fox Sports Network.

Ryan's life story epitomizes the tattoo across his fingers, "self made". He is living proof that with ambition, drive, and heart you can still capture the American dream.

William L. Galvery, Jr. was for many years Professor of Welding Technology and Welding Department coordinator at Orange Coast College in Costa Mesa, California. He graduated from California State University, Long Beach with a Bachelor of Vocational Engineering degree, has more than 30 years of industrial welding experience, and is an American Welding Society (AWS) Certified Welding Inspector and a Certified Welding Educator. In 2003 the AWS presented him with its prestigious national teaching award, the *Howard E. Adkins Memorial Instructor Award,* and for the second time he was chosen AWS District 21 Educator of the Year. In the same year he was also presented with an Excellence in Education Award by the University of Texas, Austin. Bill has also served as an officer for the AWS Long Beach/Orange County Section.

INTRODUCTION

The term welding conjures up a number of images—everything from rows of welders working among flying sparks on a factory floor to the single homeowner making repairs in the family garage. And they would both be right. Joining metals together through welding plays a large part in the manufacturing of thousands of products. It is also a tool used by do-it-yourselfers and hobbyists to create projects and make repairs around the home.

The Art of Welding is geared toward anyone who wants to learn the basics of welding for their own uses. One of the first things they will learn is that welding is really a range of processes that have evolved since the late nineteenth century. The earliest form, which is still in widespread use today, relies on a combination of ignited gases to produce a high-intensity flame that can melt metal; more modern versions use the heat produced by an electrical arc to melt and then fuse metal together.

This book covers the principal processes used by the home welder. They are oxyacetylene welding, or gas welding, and those that rely on electrical arcs, which include stick welding, MIG and TIG. Each has its own unique characteristics as well as specialized equipment. Chapters devoted to each cover equipment selection and use, tools, safety precautions, and welding techniques provided by Ryan Friedlinghaus and the pros at West Coast Customs. There are also chapters on the related skills of thermal cutting, and brazing and soldering. A final chapter offers welding tips and techniques for dealing with common problems found around the home welding shop.

AN OVERVIEW OF WELDING

The term "welding" comprises a number of different processes for fusing metals. For the home hobbyist or do-it-yourselfer, oxyacetylene welding, shielded metal arc welding (also called stick welding), wire-feed welding processes (also called MIG), and nonconsumable electrode welding (also called TIG) are the most popular. This chapter also covers the basics of joint preparation and the most common types of welds. For procedures that relate to specific welding processes, see the relevant chapter.

WHAT ARE THE MOST POPULAR WELDING PROCESSES
FOR THE DO-IT-YOURSELFER?

HOW ARE WELDING JOINTS PREPARED?

WHAT ARE SOME OF THE BASIC TYPES
OF WELDS?

WHAT ARE THE SPECIAL PROCEDURES
FOR WELDING THICK PLATES?

WHAT ARE SOME COMMON WELDING POSITIONS?

What are the most popular welding processes for the do-it-yourselfer?

Oxyacetylene Welding

When combined in the correct proportions in the welding torch handle, oxygen and acetylene gases produce an approximately 5600°F (3100°C) flame at the torch tip. This flame melts the edges of the base metals to be joined into a common pool. Sometimes additional filler metal is added to the molten pool from a welding rod. When this common pool cools and the metal freezes solid, the joined metals are fused together and the weld complete.

Oxyacetylene welding was first used industrially in the early years of the twentieth century. Although this process makes excellent welds in steel, it is little used for welding today except for a few specialties because there are other more efficient welding processes available. However, oxyacetylene has many other important uses: cutting, hardening, tempering, bending, forming, preheating, postheating, brazing, and braze welding. Because of the precise control the welder has over heat input and its high-temperature flame, together with its low equipment cost, portability, and versatility, it remains an essential tool. As with all effective tools, using oxyacetylene carries risk. We will cover the theory and use of oxyacetylene equipment so you can use them with confidence and safety. See Figure1-1.

2

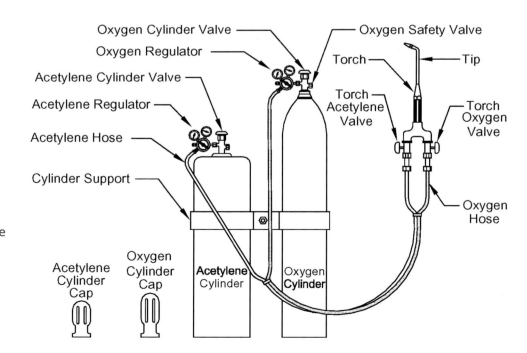

Figure 1-1
Oxyacetylene
setup.

Shielded Metal Arc Welding or Stick Welding

In shielded metal arc welding, an electric circuit is established between the welding power supply, the electrode, the welding arc, the work, the work connection, and back to the power supply. The arc produces heat to melt both the electrode metal and the base metal. Temperatures within the arc exceed 6,000°F (3,300°C). The arc heats both the electrode and the work beneath it. As the electrode moves away from the molten pool, the molten mixture of electrode and base metals solidifies and the weld is complete.

Arc welding machines have been used in this country since the early days of the twentieth century. Arc welding is popular for industrial, automotive, and farm repair because its equipment is relatively inexpensive and can be made portable. More welders have learned this process than any other. Although it will be around for many years, and its annual filler metal poundage continues to grow, it is declining in importance as wire feed welding processes continue to gain popularity and market share. We will cover theory, equipment, electrode rod classification and selection, and safety.

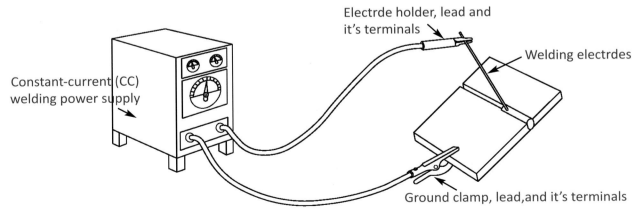

Figure 1-2. Arc welding setup

Wire Feed Welding or MIG Welding

In this process, welding wire within the welding gun is both the electrode and the filler material. Welding begins as the section of electrode wire between the tip and the base metal is heated and deposited into the weld. As the wire is consumed, the feed mechanism supplies more electrode wire at the pre-adjusted rate to maintain a steady arc.

The wire feed processes consume over 70 percent of *total* filler materials used today, and this percentage continues to grow. While this welding equipment may cost more than arc welding equipment of the same capabilities, it offers higher productivity, and it is easy to learn. Not having to stop a bead, change electrodes, and restart again increases metal deposition rates and reduces weld discontinuities. Also, these processes are readily adapted to robotic/computer-controlled operations. Wire feed processes are relatively easy to learn, especially to those already trained in shielded metal arc welding, once the power source differences and voltage- amperage variables are understood. See Figure 1-3.

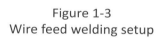

4

Figure 1-3
Wire feed welding setup

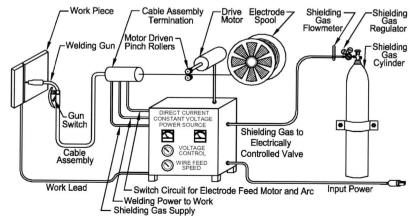

Non-Consumable Electrode or TIG Welding

A continuous arc forms between a tungsten electrode on the welding torch and the work. The electrode in this process is not consumed. However, some applications require the use of a filler rod.

Although this process requires more skill than most other processes and does not have high metal deposition rates, improvements in shielding gas mixtures, torch design, and power supply electronics have made it an indispensable tool where high quality welds are essential on aluminum, magnesium, or titanium. It can weld most metals, even dissimilar ones. See Figure 1-4.

Figure 1-4
TIG welding setup

How Are Welding Joints Prepared?

Some elements of welding are common to all types, such as joint preparation, welding terminology, and the like. They will be covered here. For specific techniques, see the chapters dealing with each of the main welding processes.

Joint preparation provides access to the joint interior. Without it the entire internal portion of the joint would not be fused or melted together making the joint weak. Remember that a properly made, full-penetration joint can carry as much load as the base metal itself, but full penetration will only occur with the correct joint preparation.

Usually, joints are prepared by flame cutting, plasma arc cutting, machining, or grinding; however, castings, forgings, shearing, stamping, and filing are also common methods used to prepare material for welding. See figure 1-5.

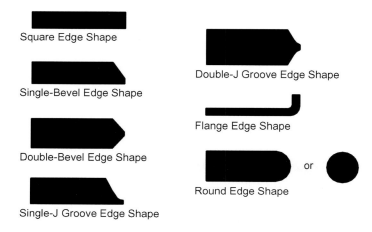

Figure 1-5 Edge shapes for weld preparation

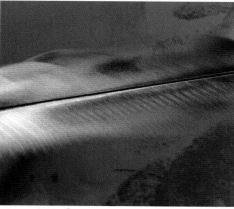

Figures 1-5 and 1-6 Proper joint preparation is essential to ensure strong welds. Here a portable grinder is used to bevel the edges of two thin sheets of metal

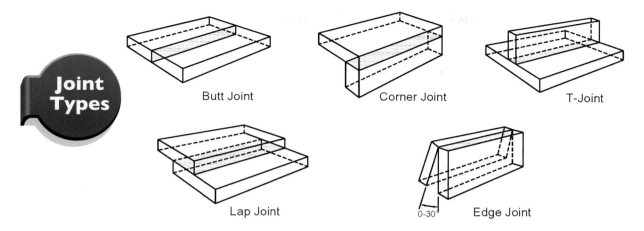

Figure 1-6 Joint types.

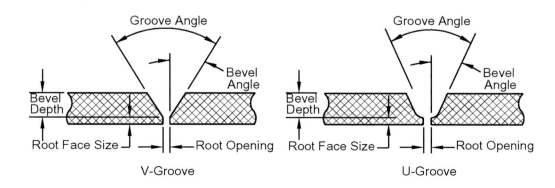

The V-and U-groove joints are common joints used in welding. The parts of the joints include

- Depth of bevel
- Size of root face
- Root opening
- Groove angle
- Bevel angle

Figure 1-7 Parts of V- and U-groove joint preparations

Joint Preparations for Butt Joints

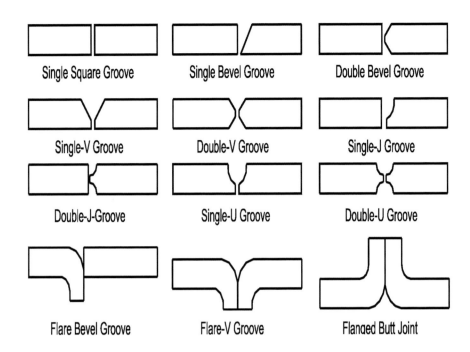

Single Square Groove | Single Bevel Groove | Double Bevel Groove

Single-V Groove | Double-V Groove | Single-J Groove

Double-J-Groove | Single-U Groove | Double-U Groove

Flare Bevel Groove | Flare-V Groove | Flanged Butt Joint

Figure 1-8 Single-groove and double-groove weld joint

Joint Preparations for Corner Joints

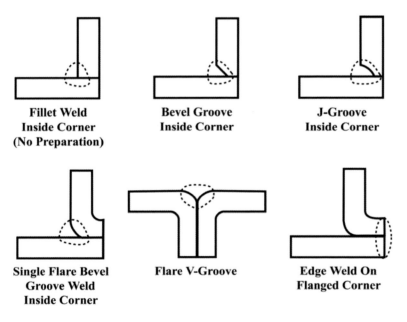

Fillet Weld Inside Corner (No Preparation) | Bevel Groove Inside Corner | J-Groove Inside Corner

Single Flare Bevel Groove Weld Inside Corner | Flare V-Groove | Edge Weld On Flanged Corner

Figure 1-9 Weld preparation for corner joints

Joint Preparations for T-joints

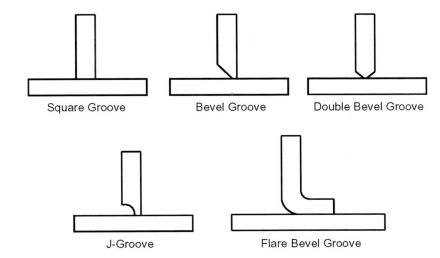

Figure 1-10 Weld preparations for T-joints

Joint Preparations for Edge Joints

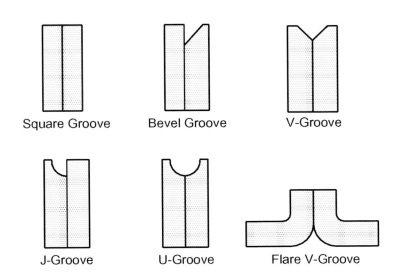

Figure1-11 Weld preparations for edge joints

Joint Preparations for Lap Joints

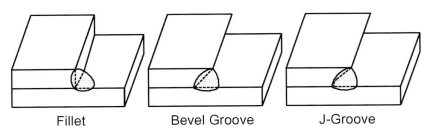

Figure1-12 Weld preparations for lap joints

9

Common Weld Preparations

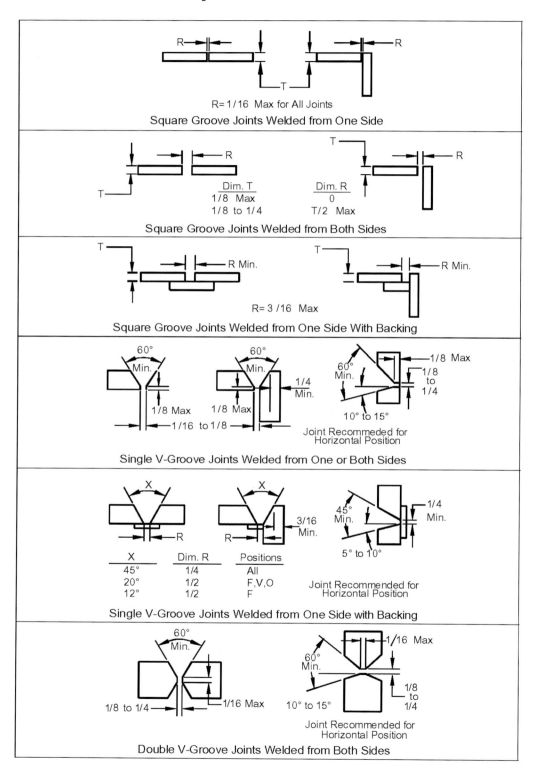

Figure 1-13 A few typical weld preparations

What are some basic types of welds?

The Groove Weld

As the name implies and the illustration below shows, this is a weld made in a groove between work pieces. See "Common Weld Preparations," on the opposite page, for some typical weld dimensions.
See Figure 1-14.

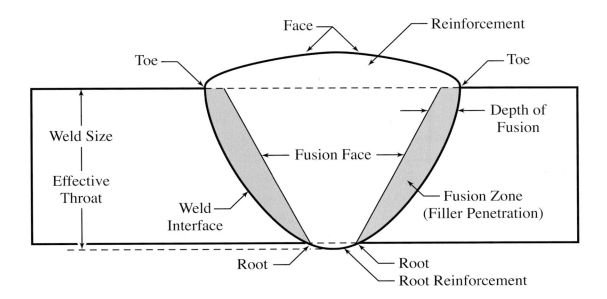

Figure 1-14 Parts of a groove weld

Fillet Weld

Fillet welds are triangular in shape and used to join materials that are at right angles to one another in a lap, T-, or corner joint. The face of the weld can be convex or concave. See Figure 1-15 A&B.

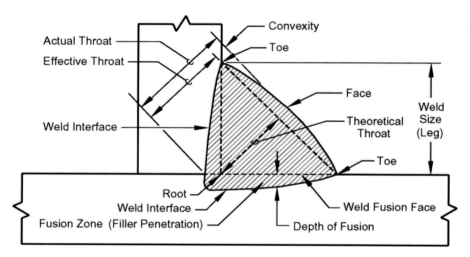

Figure 1-15A Parts of a convex fillet weld

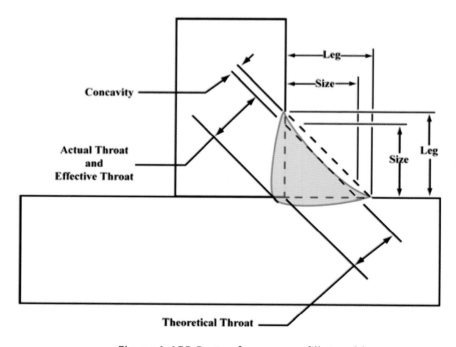

Figure 1-15B Parts of a concave fillet weld

Plug and Slot Welds

These welds join two (or more) parts together by welding them at a point *other* than their edges. They are particularly useful in sheet metal and auto-body work where welds can be completely concealed by grinding and painting. A hole or slot is made in the work-piece facing the welder and weld is made inside the hole. Filler metal completely fills the hole or slot and penetrates into the lower work-piece(s) securing them together. Plug welds are round and slot welds are elongated and rounded at the ends. See Figure 1-16.

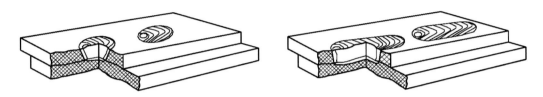

Figure 1-16 Examples of plug and slot welds

Intermittent Welds

Note the positions of the welds shown below. See Figure 1-17.

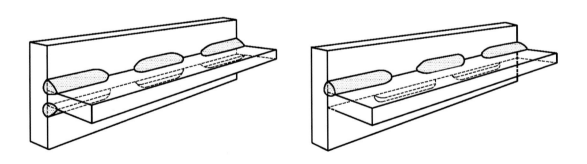

Figure 1-17 Chain intermittent fillet weld (left) and staggered intermittent fillet weld (right)

Welding Terminology

These terms describe the position of the electrode in respect to the weld.

Axis of the weld—an imaginary line drawn parallel to the weld bead through the center of the weld.

Travel angle—is the smallest angle formed between the electrode and the axis of the weld.

Work angle—for a T-joint or corner joint, the smallest angle formed by a plane, defined by the electrode (wire) and the axis of the weld, and the work piece.

Push angle during forehand welding—this is the travel angle during push welding when the electrode (wire) is pointing *toward* the direction of weld progression.

Drag angle during backhand welding—this is the travel angle during drag welding when the when the electrode (wire) is pointing *away* from the direction of weld progression.

Travel speed—the velocity or speed of the electrode (wire) along the travel axis, usually in inches/minute or cm/minute.

See Figure 1-18.

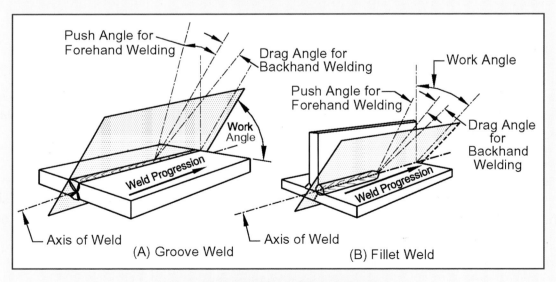

Figure 1-18 Orientation of the electrode

Tack Weld

Welders place small, initial welds along joints to hold the work pieces in place so the parts remain in alignment when they are welded. Tack welds hold work firmly in position, but can be broken with a cold chisel in the event further adjustment is needed. Beginning welders tend to make them too small. One inch is the standard length of a tack weld. A tack should be as strong as the weld itself as it becomes an integral part of the finished weld.

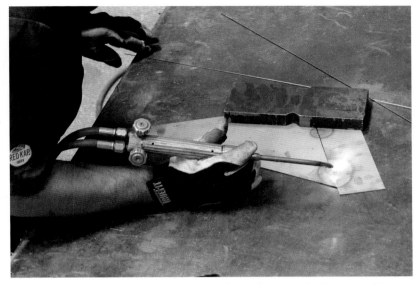

Figure 1-19 Make sure the work is aligned properly; keep it aligned by welding small sections at intervals to tack the metal in place

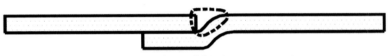

Figure 1-21 Joggle weld joint preparation

15

Joggle Joints

Joggle joints are used where a strong joint and flat surface is needed to join two pieces of sheet metal or light plate. There are hand tools available to put the joggle into sheet metal. They are useful whenever a finished surface concealing the weld is needed and where a butt joint would not work with thin sheet metal. See Figure 1-21.

Stringer and Weave Beads

In a stringer bead the path of the electrode is straight, with no appreciable side to side movement, and parallel to the axis of the weld, while a weave bead has a side-to-side motion which makes the weld bead wider (and the heat-affected zone larger) than that made with a stringer bead.

Padding

Padding is when weld filler metal is applied to a surface to build up a plate or shaft, to make a plate thicker, or to increase the diameter of a shaft. It is used either to restore a dimension to a worn part or to apply an extra hard wear surface. See Figure 1 -20A shaft, bar, or pipe and 1-20B is resurfacing a plate.

COMMON WELDING ELEMENTS

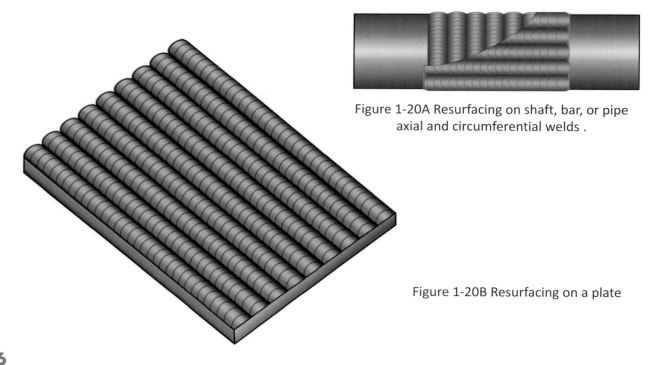

Figure 1-20A Resurfacing on shaft, bar, or pipe axial and circumferential welds .

Figure 1-20B Resurfacing on a plate

Boxing

Boxing is when a fillet weld is continued *around* a corner. Normally a fillet weld is made from one abrupt end of the joint to the other abrupt end of the joint. See Figure 1-21.

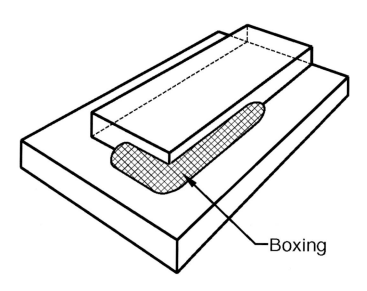

Figure 1-21 Boxing weld

Boxing

What are the special procedures for welding thick plates?

Root Pass Weld

A root pass uses weld filler metal to close the root space between the weld faces. It is especially helpful in welding pipe and thick plates where only one side of the weld is accessible and no backing material is used.

Back and Backing Welds

A back weld is applied after a groove weld is completed. The back weld is made to insure full penetration through the material being joined. Before we apply the back weld we must grind or gouge into the bottom of the groove weld until we reach sound weld metal then we may apply the back weld to the bottom of the groove weld. See Figure 1-22.

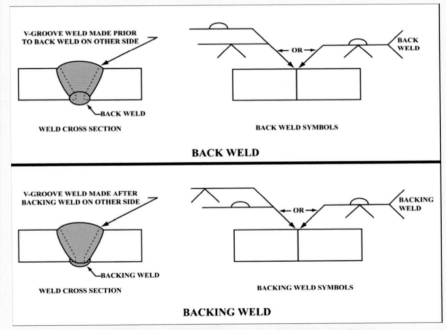

Figure 1-22 and 23 Back weld and backing weld

A backing weld is applied to the bottom or root of a groove weld before the groove weld is applied. Because the root or bottom of the weld is made first it becomes a backing for the groove. The difference between a back and backing weld is the sequence of welding. Before the groove weld is completed the backing weld must be ground or gouged to sound weld. See Figure above.

17

Backing Plates

A backing plate contains the large weld pool when joining two thick sections that are accessible from only one side. It takes the place of a root pass. The backing plate also shields the weld pool from atmospheric contamination coming in from the back of the weld. Backing plates are usually tack welded to the two sections of the weld, but there are proprietary ceramic tapes and metal-glass tapes that perform the same function and do not need to be tacked into place. Copper and other materials are also used as backing plates. See Figure 1-24.

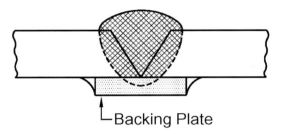

Figure 1-24 Weld backing plate

Runoff Plates

This is a made of the same material as the work being joined. The plate is tack welded to the joint at the start and/or end of the groove joint. The runoff plate contains a groove like the pieces being joined. It prevents the discontinuities caused by beginning and ending the welding process. See Figure 1-25.

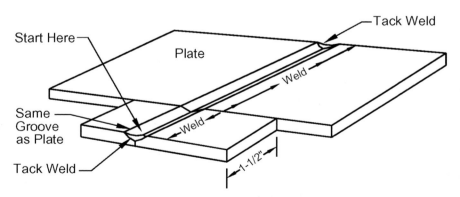

Figure 1-25 Runoff plate or tab

Multiple Pass Technique

Use multiple passes of parallel weld beads when you are faced with making a large weld but the electrode deposition is much smaller than the weld width. See Figure1-26.

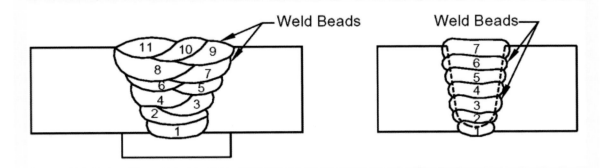

Figure 1-26 Multiple passes to join thick material

What are some common welding positions?

They are determined by the position of the axis of the weld with respect to the horizontal and whether they are made on plate or pipe. See Figure 1-27.

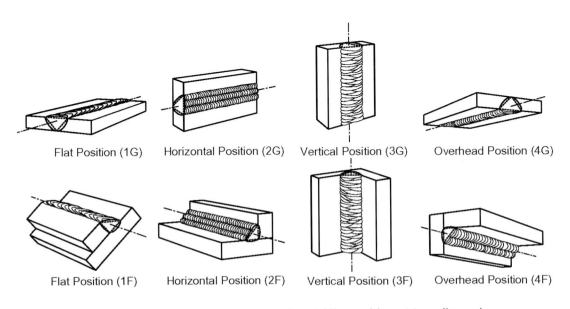

Flat Position (1G)   Horizontal Position (2G)   Vertical Position (3G)   Overhead Position (4G)

Flat Position (1F)   Horizontal Position (2F)   Vertical Position (3F)   Overhead Position (4F)

Figure1-27 Groove weld (upper) and fillet weld positions (lower)

Welding Positions for Pipe

See Figure 1-28. Note the difference between welding positions A , and C: In position A, (1G) the pipe may be rotated about its longitudinal axis to provide access to any part of the weld joint allowing the welder the opportunity to weld the entire pipe in the flat (1G) position; in position C, (5G) the pipe is fixed and *cannot* rotate forcing the welder to weld upward or downward vertically, flat on the top and overhead on the bottom; position B is pipe in a vertical position and welded on the horizontal plane; pipe in D is on a 45° angle and all positions (flat, horizontal, vertical, and overhead) are welded when pipe is in this position; the final position is pipe at a 45° with a restrictor in place (the restrictor allows the welder to weld only from one side of the restrictor) making this the most difficult of all welding positions.

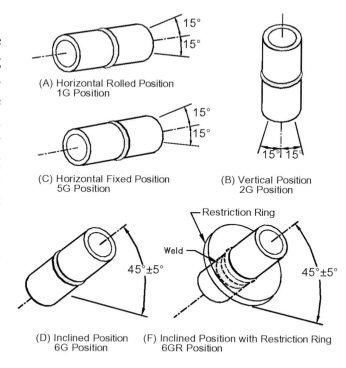

(A) Horizontal Rolled Position
1G Position

(C) Horizontal Fixed Position
5G Position

(B) Vertical Position
2G Position

(D) Inclined Position
6G Position

(F) Inclined Position with Restriction Ring
6GR Position

Restriction Ring

Weld

45°±5°

45°±5°

Figure 1-28 Pipe weld positions

Slag Removal

Remove slag between weld passes or the remaining slag will form inclusions *within* the weld metal and weaken it. Slag is usually removed with a slag hammer and wire brush angle grinders or pneumatic peening tools may also be used. Sometimes a wire wheel is used. Pipe welding, grinders and power wheels are used between each welding pass to assure a slag-free surface on which to begin the next pass.

GENERAL TOOLS, MATERIALS AND SAFETY EQUIPMENT

Each welding process requires specialized equipment to do the job. But welding, whether you use it for hobby purposes or to make home repairs, requires a number of common hand and power tools. Many of these tools you may already have in your toolbox, others you may need to purchase. One thing is certain, you will need a wide assortment of clamps to hold the work in position—both to keep the work steady while you weld and to reduce weld-induced distortion. The metals you will be working with all have different characteristics and react differently under the intense heat of the welding torch. This chapter will look at some of the properties of those metals. And, finally, you will also require specialized safety equipment.

WHAT HAND TOOLS ARE USED IN WELDING?

WHAT POWER TOOLS ARE USED IN WELDING?

WHAT TYPES OF CLAMPS ARE USED IN WELDING?

WHAT ARE THE MOST COMMON STEEL PRODUCTS USED IN WELDING?

WHAT ARE SOME BASIC SAFETY REQUIREMENTS?

What hand tools are used in welding?

A welder removes slag after a welding pass

You may already own many of these tools because they are commonly used for general carpentry and household repair. Some tools that you should have to facilitate your welding include:

- Builder's and torpedo levels—Use the larger builder's level whenever possible; it is more accurate and measures over a longer span; use the torpedo level wherever the builder's level won't fit.
- Framing, carpenter's, cabinet maker's, and combination squares—Use the largest square that fits the work. The combination square is convenient for layout of 45° corner cuts and parallel lines.
- Cold chisel and ball peen hammer—Handy for breaking tack welds when they must be repositioned; also useful for removing material between a series of drilled holes (chain drilling).
- Center Punch—Marks hole centers and cutting lines.
- Compass and dividers—For scribing circles or stepping off a series of equal intervals.
- Files—For bringing an oversized part down to exact dimension or removing a hazardous razor/burr edge.
- Hack saw—For slow, but accurate metal cutting.
- Tape measures—16- and 24-foot tapes are the most convenient sizes. Useful for measuring on curved surfaces too.
- Precision steel rules—Available in lengths from 6 to 72 inches (150 to 1000 mm).
- Protractor—For finding angles.
- Trammel points—These points fit on and adjust along a wood or metal beam and scribe circles or arcs with 20- to 40-foot (6 to 12 m) diameters. See Figure 1–2.

Figure 2-1 Here is a selection of typical hand tools used in welding, including an assortment of hammers, wrenches, screwdrivers, and pliers (top), as well different types of squares, such as carpenter's squares, speed squares, and combination squares. You will find uses for both builder's levels and smaller torpedo levels (below)

23

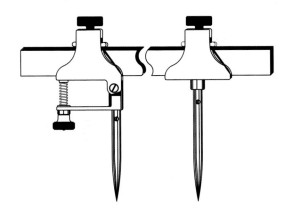

Figure 2-2 This page contains an assortment of common hand tools, including hammers, pliers, wrenches, and levels. At left are trammel points for scribing large circles

Metal Marking Tools

There are a variety of tools used to mark metal layout lines. They include

- Chalk line snap
- Weldor's chalk, also called soapstone
- Ball point metal marker
- Single center punch mark or a line of punch marks
- White lead or silver lead pencil
- Felt-tip pen
- Aerosol spray paint
- Scriber on bare metal
- Scriber on layout fluid

Use weldor's chalk for marking rough dimensions or to indicate cutting lines that will hold up under cutting torch heat. A line of center punch marks can be more accurate and will also withstand torch heat. For very accurate layout lines, spray paint the metal in the area of the layout lines and use a scriber to scratch through the paint to make the layout lines. Alternatively, machinist's layout fluid (Dykem® is the major brand, and available in red or blue) can be used to make the scribed lines more visible. These lines will not hold up under torch heat, but can be essential to laying out non-torch cutting lines. A black felt-tip pen can also be used in place of spray paint or layout fluid to darken the metal and show up scriber lines. Do not use scribe marks to designate bend/fold lines since they will be stress raisers and the part will eventually fail along the scribed line. Metal markers are available. They put down a 1/16-inch width line, come in several colors, and are excellent for applying lettering to metals. The are rated at 700°F (370°C) so cannot be used for torch cutting lines. Note that some marking materials' residues may contaminate GTAW welds.

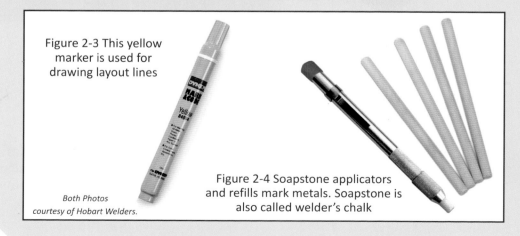

Figure 2-3 This yellow marker is used for drawing layout lines

Both Photos courtesy of Hobart Welders.

Figure 2-4 Soapstone applicators and refills mark metals. Soapstone is also called welder's chalk

24

Specialized Hand Tools

Here are some tools most welders will find necessary. They include:

- Chipping hammer to remove welding slag.
- Wire brush for cleaning welds.
- Hammer and cold chisel to break tack welds.
- Pliers for moving hot metal safely.
- Wire cutters to trim electrode wire.

Figure 2-5 Here is a group of specialized welder's hand tools
Photo courtesy of Hobart Welders.

What common power tools are used in welding?

Many of the power tools used by welders are also used for carpentry and other tasks. They include:

- Reciprocating saw—Excellent for rough cuts of bar stock, shapes, pipe, and plate.
- Hand-held band saw—Capable of accurate cuts and following scribed lines; excellent for both pipe and tubing. Its throat is too small for most plate cutting.
- Electric drill, drill bits—For starter holes to begin oxyfuel cutting, holes toinstall hardware, and for chain drilling.
- Abrasive cutoff saw—Good for rod and pipe. Not good for shapes and tubing. Difficult to make accurate cuts.
- Portable grinder with abrasive and wire wheels and abrasive flapwheels—To remove mill scale, rust, and paint before welding. It is also good for smoothing rough edges and removing bad welds.
- Bench grinder/pedestal grinder with abrasive and wire wheels—Same as portable grinder, but here the operator holds the parts.

25

Figure 2-6 Welders find that portable power saws and grinders come in handy

Figure 2-7 Bench grinders are one way to sharpen TIG electrodes before welding

What types of clamps are used in welding?

Clamps play a critical role in holding the parts to be welded in the proper position to make the weld and in preventing weld-induced distortion. Welders use a variety of general purpose clamps and clamps specifically designed for welding projects.

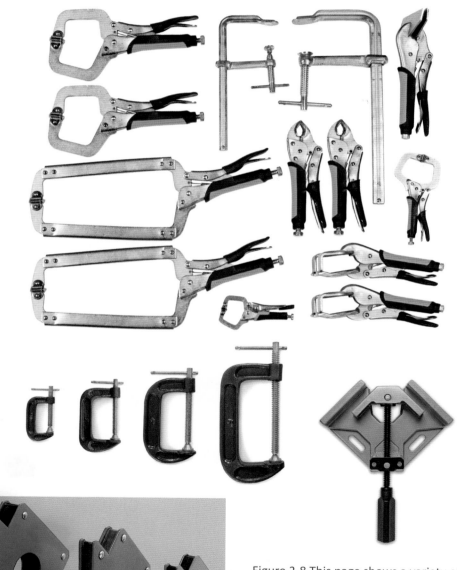

Figure 2-8 This page shows a variety of clamps used in welding, including locking clamps in various shapes and sizes and common C clamps. The clamp shown directly above is an angle clamp; to the left is a group of magnetic clamps

Photos courtesy of Hobart Welders.

Welding Tables

A welding table places the work at a comfortable height and allows weldors to concentrate on their work, rather than their discomfort. A welding table allows some welds to be made sitting down. Also, it provides a stable, flat surface to position and clamp work prior to welding. In some applications the work itself may be tack-welded to the table. Later these weld-tack beads can be ground off.

Figure 2-9 A folding, portable welding table offers

Photo courtesy of Miller Electric.

What are the most common steel products used in welding?

Low carbon, hot rolled solid shapes, sheet goods, plate, pipe, and tubing are the most often used fabrication materials. Large steel distributors stock a wide variety of shapes and sizes. Additional sizes and shapes of steel as well as other materials like alloy steels, stainless steel, brass, copper, and bronze are also available on order. These materials are substantially more expensive than carbon steel.

When planning projects, remember that one size of tubing (either round, square, or rectangular) is dimensioned to *telescope* or slide smoothly into the next larger size. This can be very helpful and a design shortcut.

Large steel distributors usually have a variety of remnant material that is sold by the pound. A tape measure and calipers can also help you determine if a particular remnant will be useful. This can be both economical and convenient for many projects. Bring work gloves to handle the greasy and sharp remnants.

Usually for a small charge, or often at no charge, the distributor will cut the material to make it easier to transport since many products come in 20 foot lengths. With good planning, cuts made on the distributor's huge shears or band saws can save you a lot of cutting time, particularly on heavy plate goods.

Fig.2-10 Common steel products used in welding projects

28

Table 2.20 Commonly Available Metal Sizes			
Product	**Maximum Thickness**	**Width or Diameter**	**Incremental dimensions**
Sheet Steel	>3/16″ thickness	36-84″	Even numbered gauges are most common: other gauges on special order.
Plate Steel	≤3/16″ thickness	>8″ to 60″	1/32″ up to 1/2″ thickness, then 1/16″ up to 1″ thickness, then 1/8″ up to 3″ thickness, and 1/4″ above 3″ thickness.
Rectangular Bars		8″	Thickness: 1/16″ increments, but usually 1/8″ increments. Width: 1/4″ increments.
Round Bars		8″	Diameter: 1/8″ increments.

Other Hardware Items

Common hardware items used by welders include

- **Nuts**—By welding a nut over an existing hole, we can add threads without tapping them; this facilitates adding leveling jacks, adjustment members, and clamps.

- **Bolts**—These can be used for leveling jacks, axels, swivel points, and locating pins. Bolts can also be cut to provide just the threaded portion for threaded studs, or just the rod section.

- **Allthread**—This is rod stock threaded end-to-end and useful where clamping or positioning action is needed.

- **Hinges**—There are three main hinge designs:
  - **Leaf hinges for welding**—These hinges are not plated and have no screw holes.
  - **Cylindrical weld hinges**—These are made in a wide variety of sizes and can support heavy loads.
  - **Piano hinges**—Provide continuous support along a door or cover.

- **Casters and Wheels**—These are better purchased than shop-made.

Table Appendix B-2 Properties of Important Metals			
Metal	**Weight (lb/ft³)**	**Melting Point**	
		°C	**°F**
Bronze (90% Cu-10% Sn)	548	850-1000	1562-1832
Brass (90% Cu-10% Zn)	540	1020-1030	1868-1886
Brass (70% Cu-30% Zn)	527	900-940	1652-1724
Bronze (90% Cu-9% Al)	480	1040	1905
Bronze, Phosphor (90% Cu-10% Sn)	551	1000	1830
Bronze, Silicon (96% Cu-3% Si)	542	1025	1880
Iron, Cast	450	1260	2300
Iron, Wrought	485	1510	2750
Steel, high-carbon	490	1374	2500
Steel, low-alloy	490	1430	2600
Steel, low-carbon	490	1483	2700
Steel, medium-carbon	490	1430	2600

Cu = Copper   Sn=Tin   Zn = Zinc   Al = Aluminum   Si =Silicon

Cleaning Metals

Some welding processes are fairly tolerant of mill scale and small amounts of rust and paint, so it is possible to make good welds on most steel rolled goods—flats and shapes—as they come from the mill. However, the metal must not be greasy and for this reason most hollow products like pipe, tubing, and hollow rectangular shapes that are shipped from the factory well oiled must be degreased before welding.

Household cleaners like Simple Green® or Formula 409® All Purpose Cleaner will work; industrial degreasers like denatured alcohol and acetone can also be used. Paint stores, metal supply houses, hardware stores, and pool supply stores often carry phosphoric acid (dilute in 4 to 10 parts water). These stores may carry tri-sodium phosphate (TSP), also a good cleaner. Do not clean hollow steel shapes too far in advance of welding or they will rust. Do not use compressed air to dry them off as this will re-introduce oil contamination from the compressor. Use a plastic bristle brush or a stainless-steel brush since copper, brass, or aluminum brushes will contaminate the weld.

Thorough Prep Work

- Grind, wire brush with a grinder, use flap wheels, or emery cloth to remove all mill scale, rust, paint, and dirt and get down to fresh metal.

- Wipe cleaned weld area with alcohol or acetone to remove residual grease.

- Avoid getting your fingerprints on the area just cleaned.

- Remember not to cross-contaminate your wire brush, emery cloth, and flap wheels by using the same ones on both steel and stainless steel. Hint: If you will be working with both steel and stainless steel, paint the handles of stainless steel brushes red for use on steel, and green for use on stainless. This prevents cross-contamination.

Figure 2-11 As a final prep step, clean the area with alcohol or acetone

Protecting Metal

Preparation for welding removes mill scale, grease, and paint, thus exposing fresh, bare metal to the atmosphere, an ideal condition for rapid corrosion. This is particularly true for most steels and for aluminum in a salt atmosphere. A protective finish will prevent corrosion and enhance the part's appearance.

The most common protective finishes for welded products include:

- Painting for all metals—No specialized equipment is needed, but spray painting may be best for complex shapes to reduce labor expense. Several coats may be needed to make the item weatherproof.
- A red Rust-oleum® brand primer and two more finish coats will provide at least five years of rust-free service outdoors. In general, products which are supplied in aerosol cans are less durable than those supplied in conventional cans.
- Powder coating for steel and aluminum—Provides a durable, professional-looking surface with many colors and surface textures available. May be nearly as inexpensive as painting for complex shapes as it is sprayed on. Holds up well outdoors.
- Anodizing for aluminum only—This coating is durable and very thin, typically from 0.5 to 6 thousandths of an inch (0.013 to 0.154 mm). Colors are available but tend to fade in the sun. Red is the least stable color, black is much more stable. A non-colored, clear anodizing is the most stable. This is not a do-it-yourself process; leave it to specialists.

Protective coatings should go on promptly after welding (or final surface prep) so the metal does not have a chance to react with the atmosphere. Ideally only a few hours should elapse.

31

Figure 2-12 After welding, metal should be protected with paint or some other type of coating

What are some basic safety requirements?

Safe Welding Areas

Some welding processes have specific safety requirements. But in general the welding area should

- Be clean and comfortable to work in.

- Allow you to position the work to avoid welding on the floor unless absolutely necessary; you will not do you best work there.

- Be free of drafts on the work from fans, wind, windows, and doors, yet still have adequate change of air ventilation to reduce weld fume inhalation.

- Provide bright light; welding in sunlight is better than in dim light as the non-glowing parts of the weld show up better.

- Be between 70 to 80°F (21 to 27°C) because better welds will result than welds made in cold temperatures; however, acceptable welds can be made at ambient temperatures in the 40 to 50°F (4 to 10°C) range except where the weld specifications call for preheating.

- Have tools positioned within easy reach of the weldor.

- Be clear of combustibles, puddles, and tripping hazards.

- Provide all necessary personal safety equipment for the processes to be used.

Figure 2-13 A welding fume extractor removes potentially harmful vapors from the welding area without contaminating the weld

Photo courtesy of Lincoln Electric .

Personal Safety

In addition to providing a safe working environment, the weldor should also take steps toward personal protection for themselves or anyone else in the welding area. Those steps include:

- Protection of face and eyes from sparks and radiation with a helmet and lens of appropriate shade number (darkness).

- Protection of all of the welders skin from arc and weld material radiation by covering it with cotton, wool, specially treated canvas jackets, or leather garments; ultra violet radiation is carcinogenic.

- Personnel in the welding area must be protected from the welding arc and sparks by protective screens. Never view the welding being performed through the protective screens alone; the only way to safely view welding is through the proper shade lens and welding helmet or goggles.

- Beware of hazards from gases and ensure adequate ventilation; inert shielding gases may cause suffocation in confined areas.

- Provide adequate ventilation from welding process smoke and the metal vapors, particularly heavy metals like zinc and cadmium that are toxic; keep your head out of the welding plume.

- Leathers or specially treated canvas jackets must be worn when welding vertically or overhead to protect the welder from the falling hot metal, sparks and slag.

- A welder's hat will prevent both radiation burns to the head and hot sparks, falling slag, and hot metal burns.

- High-top boots can prevent hot sparks and slag from burning your feet.

- Never weld with pant cuffs; sparks falling into cuffs will burn pants.

- Make sure your welding gloves are dry and have no holes.

- Keep hands and body insulated from both the work and the metal electrode holder.

- Do not change the polarity switch position while the machine is under welding current load.

- Welding machines must be turned off when not attended.

- Do not stand on a wet surface when welding to prevent electric shock.

- Welding cables and electrode holders must be inspected for broken insulation regularly to prevent electric shock.

- Welding power supplies on AC lines must be properly grounded and emergency shut-off switch location known and accessible.

- Welding area must be dry and free of flammable materials.

- Protect your ears from welding and grinding noise with ear plugs or ear protectors.

- Any compressed gas cylinders must be properly secured and out of the spark stream.

- Avoid wrapping welding cable around your arm or body in case a vehicle snags the cables.

- Never cut or weld on containers without taking precautions.

- In shielded metal arc welding, the welder must plan for disposal of electrode stubs: they are hot enough to cause burns and to start fires and must not be dropped from heights because of the hazard to others.

33

Basic Safety Equipment

✓ Welding helmet with the proper lens shade for the process and amperage.

✓ Leather capes and sleeves or jacket called *skins* or *leathers*, to protect the welder's clothing from sparks, especially while welding overhead.

✓ Welder's cap to protect from sparks getting behind the welding helmet and into the welder's hair.

✓ Welding safety equipment

✓ Breathing apparatus to provide the welder with fresh air in confined spaces with inadequate ventilation. Safety glasses *under* the welding helmet.

Figure 2-14 Here's a sample of safety equipment. From top left going clockwise: goggles worn under welding helmet; welding caps to be worn under helmet; welding jacket; and gloves

Photo courtesy of Lincoln Electric and Hobart Welders.

Table 2-3 Recommended lens shade by amperage and welding process

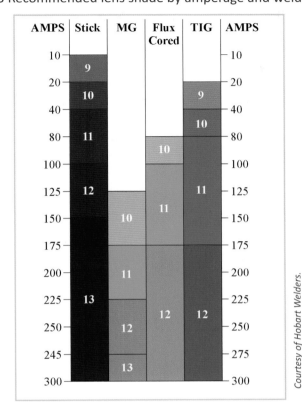

Courtesy of Hobart Welders.

Auto-darkening Faceplates

An electronic faceplate or lens is one of the most recent and important safety devices developed in the welding industry. These devices are designed to be clear, or nearly clear, then darken the instant arc is established; when purchasing be aware of the time the lens takes to darken 1/25,000 of a second or faster is recommended. Advantages of auto-darkening faceplates include:

Figure 2-15 Auto-darkening faceplates darken almost immediately as soon as the arc is established
Photo courtesy of Hobart Welders.

• The welder does not need to raise and lower his helmet when performing a series of welds: he can always see where he is with the helmet down.

• The beginner does not have to master holding his electrode steady when he drops his helmet. This permits beginners to perform better welds earlier in their training.

• Electronic faceplates offer continuous eye protection from infrared radiation coming off red-hot metal even when they are not in the darkened mode. It is just easier on the eyes and the welder is less likely to incur eye injury from inadvertent arc strikes.

OXYACETYLENE WELDING

Oxygen and acetylene gases when combined, in the proper proportions in the torch handle's mixing chamber, produce an approximately 5,600°F (3,100°C) flame at the torch tip. This flame melts the edges of the base metals to be joined into a common pool. Sometimes additional filler metal is added to the molten pool from a welding rod. When this common pool cools and the metal freezes solid, the joined metals are fused together and the weld complete. Other fuel gasses can be used in place of acetylene—and often are for soldering or brazing—but their maximum heat potential is below the heat output required for welding steel. The American Welding Federation (AWF) uses the abbreviation OAW for oxyacetylene welding.

WHAT ARE THE APPLICATIONS THAT OAW IS BEST SUITED FOR?

WHAT METALS CAN THE OAW PROCESS READILY WELD?

WHAT ARE SOME BASIC SAFETY PROCEDURES?

WHAT TYPES OF EQUIPMENT ARE NEEDED?

HOW DO YOU SELECT FILLER METAL (WELDING ROD)?

HOW IS THE EQUIPMENT SET UP?

WHAT ARE THE SPECIAL PREPARATIONS FOR WELDING?

HOW ARE WELD RESULTS ACCESSED?

WEST COAST CUSTOMS WELDING TIPS

OXYACETYLENE WELDING TIPS FROM WEST COAST CUSTOMS

Figure 3-2 OAW does not require electrical power, so it is extremely portable.

What are the applications that OAW is best suited for?

If you have many different types of projects or repairs, OAW processes work well, particularly if you are not near an electrical power source. Unlike other welding processes, OAW does not require electrical power. Use it for welding thin sheet, tubing, and small-diameter pipe.

Pros and Cons of OAW Process

Advantages

- Low cost
- Readily portable
- Excellent control of heat input and puddle viscosity
- No external power required
- Good control of bead size and shape
- Fuel mixture is hot enough to melt steel

Disadvantages

- Not economical to weld thick pieces of metal compared with other processes
- Slowest of the welding processes

What metals can the OAW process readily weld?

- Copper
- Bronze
- Lead
- Low alloy steels
- Wrought Iron
- Cast steel

Aluminum and Stainless Steel

These metals are usually not welded using the OAW process, but they may be welded provided one or more of the following steps are taken: preheat, postheat, use of fluxes, or special welding techniques. The reason is that aluminum does not change color prior to melting, so it requires extra welder skill to control heat input. It lacks strength at high temperatures. And exposed aluminum has a very thin oxide layer that requires the use of flux and also the oxide surface does not let the welder see a wet-looking molten weld pool.

What are some basic safety procedures?

All construction projects require a group of safe practices, and welding is no exception. Chapter 1 covered general safety measures for welding, but there are additional practices and equipment that are specific to OAW processes.

39

Essential Pieces of Safety Equipment

- Non-synthetic fabric (cotton or wool) long-sleeved shirt buttoned to the top to prevent sparks from entering.
- Tinted welding goggles with minimum of number 5 shade lenses.
- Leather gloves.
- Spark igniter.
- Pliers for moving hot metal.

Figure 3-3 Cotton welding shirt.

Photo courtesy of Hobart Welders.

Figure 3-2 Leather gloves.

Photo courtesy of Hobart Welders.

Figure 3-4 Tinted welding goggles.

Photo courtesy of Hobart Welders.

Figure 3-5 Spark igniter for lighting welding torch.

Photo courtesy of Hobart Welders.

Preventing Accidents

- External eye injuries from welding or grinding sparks are prevented by using welding goggles, safety glasses, or safety shields.
- Internal (retinal) eye damage from viewing hot metal and the radiation being emitted during welding and while cooling (until the metal is no longer red), prevented by using a number 5 tinted lens.
- Burns from weld sparks and hot metal prevented by leather or heavy cotton welding gloves, fire retardant clothing, leathers or specially treated welding jacket or cape-sleeves and bibs when working overhead, cuffless pants, high-top leather shoes.
- Fume hazards from the vapors of metals and flux must be avoided by proper ventilation, fume filters, and welder air supplies to the welding hood.
- Fires from the welding process prevented by moving flammables away from the weld zone and having water or fire extinguishers close at hand.

16-Step Safety Plan

40

- Never use oxygen in place of compressed air.
- Never use oxygen for starting engines or cleaning clothing.
- Store and use acetylene and propane cylinders valve end up.
- Secure cylinders to prevent them from being knocked over in use.
- Use valve protection caps on cylinders while moving them.
- Use a striker to light an oxyfuel torch. Never use a match or cigarette lighter because these can can cause a large fire or explosion with the potential power of a half-stick of dynamite.
- Never leave a lighted torch unattended.
- When a cylinder is empty, close the valve and mark it *EMPTY (MT)*.
- Do not attempt repair of cylinder valves or regulators; send them to a qualified repair shop.
- Never use compressed gas cylinders as rollers.
- Never attempt welding on a compressed gas cylinder.
- Keep power and welding cables away from compressed gas cylinders.
- Prevent sparks from falling on other persons, combustible materials, or falling through cracks in the floor.
- On old-style acetylene cylinders with a removable valve wrench, always leave the wrench in place when using the equipment so that it can be shut off quickly in an emergency.
- When transporting compressed gas cylinders by vehicle have the cylinder caps in place and secure the cylinders so they will not move around as the vehicle starts and stops. Never transport cylinders with the regulators in place.
- Never carry compressed gas cylinders inside a car or car trunk.

Welding Sealed Cylinders or Other Containers

Never weld on a sealed container regardless of its size. Even if the vessel is clean and empty, penetration of the shell could release hot gases from the interior. They could also drive the torch flame back towards the welder. If the cylinder is empty and contains no residual vapors, vent it to atmosphere by opening a valve, hatch, bung, or by drilling a hole. An even more dangerous situation results when the cylinder contains residual flammable vapors whether it is vented to atmosphere or not. This will almost certainly result in an explosion. Clean or purge the cylinder with an inert gas, then have it checked for explosive vapors by a qualified person. Vent it to the atmosphere and begin welding. In some cases filling the vessel with water or other liquid and welding below the liquid is acceptable, but this is an area for experienced, knowledgeable welders.

Ventilation

Make sure the welding area is well ventilated to draw the weld fumes away from the welder. Many fumes from the welding process are poisonous and must be avoided. Welding fumes from cadmium plating, galvanized sheet metal, lead, brass (which contains zinc), and many fluxes (especially those containing fluorine) are poisonous. They can have both immediate and long-term adverse health effects. Welding supply companies, welding equipment manufacturers, and materials suppliers will provide MSDSs (Material Safety Data Sheets) on request. Often they are available for downloading via the Internet from the manufacturer. They detail the hazards of materials and equipment and show how to deal with them safely. They are particularly helpful in understanding the fume hazards of fluxes, solders, and brazing materials.

What types of equipment are needed?

Compressed Gas Cylinders

Both oxygen and acetylene are shipped and stored in special cylinders. Oxygen cylinders are seamless vessels of special high-strength alloy steel. They are made from a single billet by a draw-forming process and they contain no welds. Acetylene cylinders are fabricated and contain welds.

Frequently, oxygen cylinders are painted green or have a green band, but the only sure

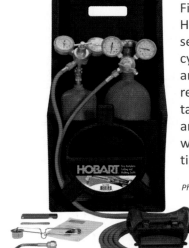

Figure 3-6
Here is a basic OAW setup. It includes cylinders of oxygen and acetylene, regulators for the tanks, hoses, igniter, and welding torch with interchangeable tips.

Photo courtesy of Hobart Welders.

way to determine the contents of a compressed gas cylinder is to *read* the adhesive label on it. This label is required by law and should not be removed. Do not go by its color as there is no color code. Unlike civilian industry, the US armed forces *do* color code their cylinders.

Acetylene will form explosive mixtures with air at *all* concentrations between 2.5 and 80%. This is the widest range of any common gas and almost insures an explosion if leaking gas is ignited. At 70°F (21°C), the acetylene should show 225 psi (15.5 bar) and the oxygen 2,250 psi (155 bar). Note that these pressures will fluctuate with ambient temperature.

Acetylene, like most other fuel gas handling equipment, has a notch or groove cut in the middle of the edges of the hexagonal faces of the swivel nut. This is a flag for a left-handed thread. See Figure 3–7.

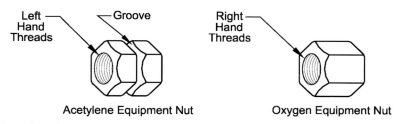

Figure 3-7 This illustration compares connector nuts for acetylene and oxygen equipment.

Tank Cross Sections

Safety valves and plugs prevent the cylinder bursting from overpressure when it is heated. Oxygen cylinders have a small metal diaphragm in a section of the valve which ruptures, releasing cylinder pressure to the atmosphere and preventing a cylinder burst. Disk rupture occurs above 3,360 psi (232 bar), the cylinder test pressure.

Acetylene cylinders contain one to four fusible safety plugs depending on their capacity. These fusible plugs, made of a special metal alloy, melt at 212°F (100°C).

They also release the cylinder contents to atmosphere to prevent rupturing (and then exploding) when the cylinder is exposed to excessive temperatures, usually from a fire. Acetylene cylinders may have the plugs on the top, or top and bottom.

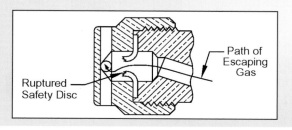

Figure 3-8 Detail of pressure safety relief on oxygen valve

Transporting Cylinders. If an acetylene cylinder has been incorrectly transported on its side, the welder avoid using it immediately. The acetylene gas and the acetone in which it is dissolved may become mixed in the area just below the valve, resulting in both gaseous acetylene and liquid acetone at the top of the cylinder. This is where acetylene exits the cylinder and goes through the valve to enter the regulator. Both acetylene gas and liquid acetone will be drawn into the regulator possibly ruining the rubber components of the regulator and torch and creating a safety hazard. The weld metallurgy may also be contaminated.

To use the cylinder, stand it upright and wait at least one-half hour before connecting and using the cylinder to allow the liquid phase of the acetone to separate from the acetylene gas in the upper portion of the cylinder. That way no acetone will be drawn into the regulator possibly damaging its seals. Also, acetone in the weld flame will contaminate the weld pool and spoil the weld.

43

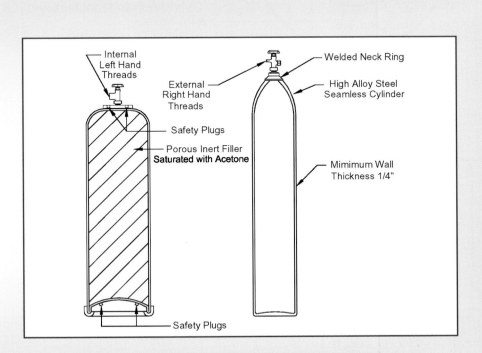

Figure 3-9 Cross sections of an oxygen and acetylene cylinders

Reading High-Pressure Cylinders

The stampings indicate which US Department of Transportation specifications the cylinder meets, what type steel was used, who fabricated it, and when.

- Steel stamp markings such as "DOT-3A-2400" indicate the cylinder was made to US Government Department of Transportation (DOT) specifications, the "3A" denotes chrome manganese steel (or "AA" for molybdenum steel), and the "2400" the maximum filling pressure in psi.

The oldest date indicates the month and year of manufacture. Subsequent dates, usually at five year intervals, indicate when mandatory hydrostatic pressure testing was performed and by whom. See Figures 3-10 and 3-11.

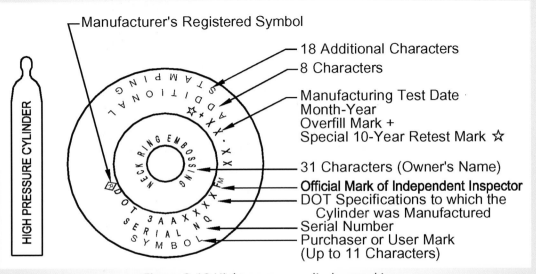

Figure 3-10 High-pressure cylinder markings

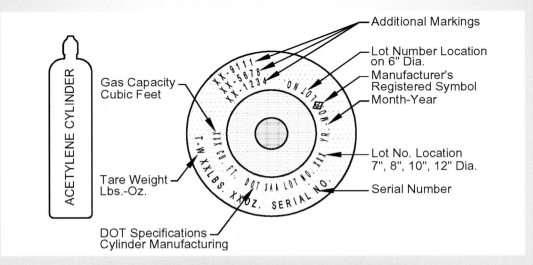

Figure 3-11 Acetylene cylinder markings

Common Cylinder Sizes

A 55 ft^3 (1557 liter) oxygen cylinder would last under two hours cutting 1/8 inch (3 mm) steel plate. For the larger cylinders, their size and weight can be major draw-backs. In general the mid-sized cylinders offer the best compromise of economy and convenience.

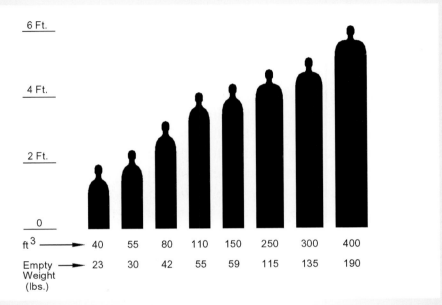

Figure 3-12 Oxygen cylinder sizes

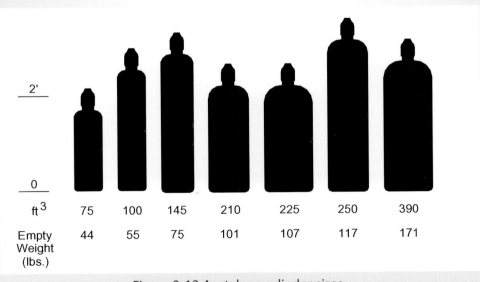

Figure 3-13 Acetylene cylinder sizes

Other Fuel Gases can be used in place of acetylene, but their maximum heat potential is below that required for welding steel. Acetylene is the best gas for welding because it:

• Has the highest temperature of all fuel gases.
• Acetylene delivers a higher concentration of heat than other fuel gases.
• Has the lowest chemical interaction with the weld pool's molten metal than all other gases.

However, other gases such as natural gas, methylacetylene-propradene stabilized (also called MPS or MAPP® gas), propane, hydrogen, and proprietary gases based on mixtures of these are frequently used for other non-welding processes for cost reasons. They work well for soldering, brazing, preheating, and oxygen cutting, and are seldom used for welding. Small changes, like different torch tips, may be necessary to accommodate alternate fuel gases. Table 3–1 shows the maximum temperature achievable with different fuel gases. Where even lower temperatures are needed (sweating copper tubing and many small soldering tasks) a single cylinder of fuel gas using only atmospheric oxygen is effective and economical.

Table 1.1 Combustion properties of fuel gases.

Fuel Gas	Oxygen-to-Fuel Gas Combustion Ratio	Neutral Flame Temperature with Oxygen °F	°C	Total Heat Btu/ft^3	Specific Gravity (Air = 1.0)
Acetylene	2.5	5589	3087	1470	0.906
Propane	5.0	4579	2526	2498	1.52
Methylacetylene-Propadiene (MPS, MAPP®)	4.0	5301	2927	2460	1.48
Propylene	4.5	5250	2900	2400	1.48
Natural Gas	2.0	4600	2538	1000	0.60
Hydrogen	0.5	4820	2660	325	0.07

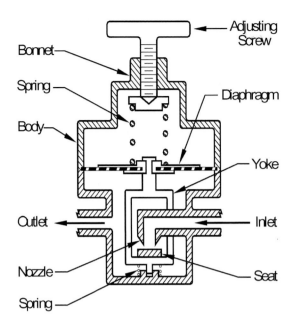

Figure 3-14 Two-stage regulator

Regulators

Regulators reduce the pressures of welding gases from the very high cylinder pressures to the low pressures needed by the torch to function properly. Also, as the cylinder pressure falls with gas consumption, the regulator maintains the constant pressure needed by the torch, even though the cylinder supply pressure drops greatly. For example, an oxygen cylinder may contain oxygen at 2,250 psi (155 bar) and the torch requires about 6 psi (0.4 bar) to operate. Similarly, a full acetylene tank may contain gas at 225 psi (15.5 bar) and the torch needs fuel gas at 6 psi (0.4 bar).

There are single-stage and two-stage regulators available. The two-stage regulator's advantage is that a higher volume of gas may be withdrawn from the cylinder with less pressure fluctuation than produced by a single-stage regulator. The combination of two regulators working together in series maintains a very constant torch pressure over wide cylinder pressure changes. Its disadvantage is cost. They are only needed when large gas volumes are needed as with multiple stations or rosebud tips.

Figure 3-15 This is an oxygen tank regulator

Torches, Tips, and Hoses

Shown below is the most common oxyacetylene torch design. Other designs are available. Some have very small flames for jewelry and instrument work, while others take no accessories and are much lighter in weight than standard torch designs to reduce operator fatigue. Some torch handles can accommodate cutting heads.

Matching the size of the flame, which is controlled by the torch tip, and the resulting volume of gas to the thickness of the metal in the weld is important. Too much flame and the base metal around the weld may be damaged, too little and there is inadequate heat to melt metal for full penetration.

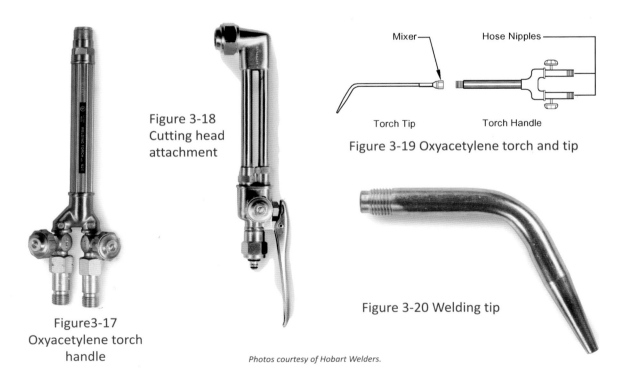

Figure 3-16 Tank hoses are sold in 25 and 50-foot lengths

Figure 3-18
Cutting head
attachment

Figure 3-19 Oxyacetylene torch and tip

Mixer

Hose Nipples

Torch Tip

Torch Handle

Figure3-17
Oxyacetylene torch
handle

Figure 3-20 Welding tip

Photos courtesy of Hobart Welders.

Torch Tip Sizes

There is no industry standard; each torch manufacturer has its own numbering system. Cross-reference tables compare each manufacturer's tip sizes with numbered drill sizes.

The American Welding Society (AWS) has been urging tip manufacturers to stamp tips with the material thickness size to eliminate the confusion of tip size numbers. The AWS C4.5M Uniform Designation System for Oxy-Fuel Nozzles calls for tips to be stamped with the name of the manufacturer, a symbol to identify the fuel gas, the maximum material thickness, and a code or part number to reference the manufacturer's operating data; many manufacturers are not in compliance. Most companies making welding tips do provide information booklets available to cross reference their tip sizes to tip drill sizes. See Table 3-2.

Table 3-2 Matching welding tip size to weld material thickness

Material Thickness Range (in.)	Filler Rod Diameter (in.)	Tip Drill Size	Orifice Size (in.)	Approximate Flame Cone Length
22-16 gauge	1/16	69	0.029	3/16
1/16-1/8	3/32	64	0.036	1/4
1/8-3/16	1/8	57	0.043	5/16
3/16-5/16	1/8	55	0.052	3/8
5/16-7/16	5/32	52	0.064	7/16
7/16-1/2	3/16	49	0.073	1/2
1/2-3/4	3/16	45	0.082	1/2
3/4-1	1/4	42	0.094	9/16
Over 1	1/4	36	0.107	5/8
Heavy Duty	1/4	28	0.140	3/4

49

Finding the Drill Size of a Tip

Using a tip cleaner find the round file which fits into the tip easily but snuggly then check the drill size of that file listed on the body of the tip cleaner cover.

TOOL TIP Cleaning Tips

When sparks from the weld puddle deposit carbon *inside* the nozzle and on the tip face, they act as spark plugs and cause premature ignition of the gas mixture. Torch tips should be cleaned at the start of each day's welding and whenever flashback occurs, the flame splits, or when the sharp inner cone no longer exists. To clean, select the largest torch tip cleaning wire file that fits easily into the nozzle and use the serrated portion to remove any foreign material. Be careful not to bend the tip cleaner file into the tip which can cause the cleaning file to break inside the tip; if the file breaks inside the tip it is nearly impossible to remove. Also be sure not to enlarge the existing hole. Then touch up the face of the tip with a file or emery cloth to remove any adhering dirt. Use compressed air or oxygen to blow out the tip. Never use a twist drill to clean the tip; it will cause bell-mouthing.

Figure 3-22 Here is a collection of torch cleaning files.

Photo courtesy of Hobart Welders.

Flashback and Backfire

Flashback occurs when a mixture of fuel and oxygen burns inside the mixing chamber in the torch handle and reaches the hoses to the regulators or cylinders. *Such burning in the hoses is extremely dangerous and will lead to serious injury.* If either through operator horseplay (like turning on both the acetylene and the oxygen with the torch tip blocked) or through regulator failure, an explosive mixture of acetylene and oxygen is forced back toward the cylinders. When the torch is lit, this explosive mixture will go off. See Figure 3-23.

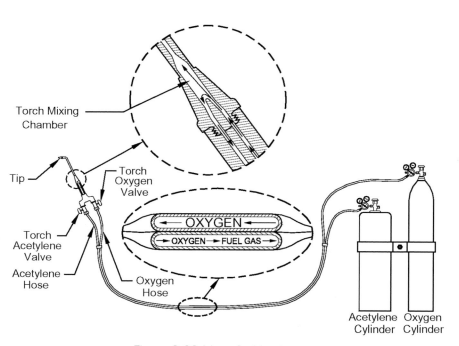

Figure 3-23 How flashback can occur

Flashback is easily prevented by installation of flashback arrestors consisting of both a check valve and a flame arrestor. These devices are about the diameter of the gas hoses and about 1 3/4 inches long. Some newer torch designs incorporate check valves and flashback arrestors into the torch handle itself. Some arrestors fit between the regulator and the hose. See Figure 3-24. The best arrestors include a thermally-activated, spring-loaded shut-off valve which closes on sensing a fire.

A **backfire** is a small explosion of the flame at the torch tip. The biggest hazard is that the detonation from the tip may blow molten weld metal five to ten feet from the weld and injure someone. Also, a series of repeated, sustained backfires, which can sound like a machine gun, may overheat the tip or torch, permanently damaging them.

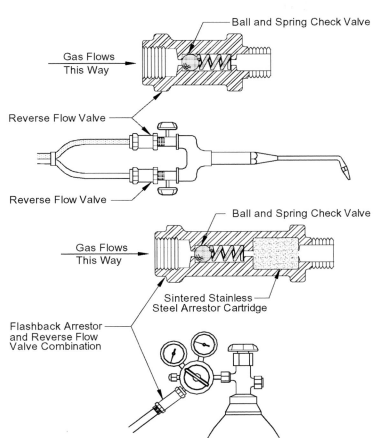

Ball and Spring Check Valve

Gas Flows
This Way

Reverse Flow Valve

Reverse Flow Valve

Ball and Spring Check Valve

Gas Flows
This Way

Sintered Stainless
Steel Arrestor Cartridge

Flashback Arrestor
and Reverse Flow
Valve Combination

Figure 3-23 Reverse-flow check valve flashback
arrestor cross section

Figure 3-25 The flashback
arrestor on the right is for
an oxygen line; the one on
the left is for acetylene.
Photo courtesy of
Hobart Welders.

51

The most frequent cause of backfire is pre-ignition of the mixed acetylene and oxygen. Here are the most common causes of pre-ignition and their solutions:

• The mixed welding gases are flowing out through the tip more slowly than the flame front burns and the flame front ignites the gas in the tip and/or mixing chamber causing a pop. Solution: Slightly increase both the oxygen and acetylene pressures and if this results in too large a flame for the job, reduce the torch tip size.

• The tip may be overheated from being held too close to the weld or from working in a confined area like a corner. Solution: Let the tip cool off and try again holding the tip farther from the weld pool.

• Carbon deposits or metal particles inside the tip act like spark plugs prematurely igniting the mixed gases. Solution: Let the tip cool, then clean it thoroughly with your tip cleaning kit.

How do you select filler metal (welding rod)?

Usually the filler metal is a close match to the base metal. Sometimes the filler metal will have deoxidizers added which will improve the weld more than just a base metal match. Rod diameters vary from 1/16 to 3/8 inch diameter. The prefix R in the description of the oxy-acetylene welding wire means rod which is followed by two or three numbers designating the ultimate tensile strength of the as welded filler material in thousands of pounds per square inch (psi). See Table 3–3.

Table 3–3 Oxyacetylene steel welding rods

AWS Classification	Minimum Tensile Strength (ksi)	Elongation in 1 inch (minimum %)
R45	—	—
R60	60	20
R65	65	16
R100	100	14

Procedures and Material Selections

Before welding, remove all surface dirt, scale, oxide, grease, and paint. Refer to Table 3-4 for suggestions on welding common metals. See Figure 3-25.

Figure 3-25 Before welding, clean metal thoroughly to remove grime and mill scale.

Table 3–4 Information for welding various metals

Metal Welded	Technique and Potential Problems	Flux Used	Flame Type	Suggested Rod
Aluminum	Al does not show color change before melting and has poor hot strength. Tack joint before welding. Remove all flux after welding.	Al flux	SR	Match base metal
Brass	Braze	Borax	SO	Navy brass
Bronze	Braze	Borax	SO	Copper-tin
Copper	Braze	—	N	Copper
Iron, Grey Cast	Pre-heat to avoid cracking. Weld at dull red heat. Flux applied to rod by dipping hot rod flux. Allow joint to cool slowly or it will crack.	—	N	Copper
Iron, Malleable Cast	Welds to poor strength. Better to braze weld using bronze rods.	Borax	—	Bronze
Iron, wrought	Weld or braze	—	N	Steel
Steel, Low-Carbon	Weld or braze	—	N	Steel
Steel, Medium-Carbon	Weld or braze	—	SR	Steel
Steel, High-Carbon	Weld or braze	—	R	Steel
Steel, Low-Alloy	Weld or braze	—	R	Steel
Steel, Stainless	Weld or braze	SS flux	SR	Match base Metal

SR = Slightly Reducing   SO = Slightly Oxidizing   N = Neutral

Figure 3-26 Be sure you have the right type flame for the job at hand. For additional information on torch flames see pages 55 and 56.

How is the Equipment Set Up?

Here are the steps for beginning the OAW process.

- Put on your welding safety equipment: tinted safety goggles (or tinted face shield), cotton or wool shirt and pants, high-top shoes, and welding gloves at a minimum.
- Make sure the valves on previously used or empty cylinders are fully closed and their valve protection covers are securely screwed in place. Then remove the empty cylinders from the work area and secure them against tipping during the wait for a refill shipment. Secure the newly replaced or full cylinders to a welding cart, building column, or other solid anchor to prevent the cylinders from tipping over during storage or use.
- Momentarily open each cylinder valve to the atmosphere and reclose the valve quickly purging the valve; this is known as cracking a valve. Cracking serves to blow out dust and grit from the valve port and to prevent debris from entering the regulators and torch.
- With a clean, oil-free cloth, wipe off the cylinder valve-to-regulator fittings on both cylinders to remove dirt and grit from the fittings' connection faces and from the fittings' threads. Do the same to both regulators' threads and faces. Remember, never use oil on high-pressure gas fittings. Oxygen at high pressures can accelerate combustion of oil into an explosion.
- Make sure reverse-flow check valves are installed on the torch or the regulators.
- Check to see that both the oxygen and acetylene regulator pressure adjustment screws are unscrewed, followed by threading each regulator to its respective cylinder. Snug up the connections with a wrench. Caution: Oxygen cylinder-to-regulator threads are right-handed; so are oxygen hose to-torch screw fittings. Acetylene cylinder-to-regulator fittings and acetylene hose-to-torch fittings threads are left-handed.
- Stand so the cylinders are between you and the regulators, S-L-O-W-L-Y open the oxygen cylinder valves. Open the oxygen cylinder valve until it hits the upper valve stop and will turn no further. Also standing so the cylinders are between you and the regulator, open the acetylene cylinder valve gradually and not more than 1 1/2 turns. If there is an old-style removable wrench on the acetylene cylinder, keep it on the valve in case you must close it in an emergency. Standing so that the cylinders are between you and the regulators offers some protection should the regulator fail and the housing and gauges explode.

54

Figure 3- 27 With the tank valves closed, back out the pressure adjusting screws on the regulators. When opening the tank valves, stand to the side or behind the tank in case the regulator malfunctions and explodes.

- Look at the high-pressure—cylinder side—pressure gauges to indicate about 225 psi (15.5 bar) in the acetylene cylinder and 2,250 psi (155 bar) in the oxygen cylinder. Note: 1 bar = 1 atmosphere = 14.5 psi = 0.1 MPa. Cylinder pressures vary with ambient temperature. The pressures given above are for full cylinders at 70°F (21°C).

- Purge each torch hose of air separately: Open the oxygen valve on the torch about three-quarters of a turn, then screw in the pressure control screw on the oxygen regulator to your initial pressure setting—about 6 psi (0.4 bar). After several seconds, close the torch valve. Do the same for the acetylene hose. Comment: We do this for two reasons, (1) to make sure we are lighting the torch on just oxygen and acetylene, not air, and (2) to set the regulators for the correct pressure while the gas is flowing through them.

  Caution: never adjust the acetylene regulator pressure above 15 psi (1 bar) as an explosive disassociation of the acetylene could occur.

- Recheck the low-pressure gauge pressures to make sure the working pressures are not rising. If the working pressure rises, it means the regulator is leaking. Immediately shut down the cylinders at the cylinder valves as continued leaking could lead to a regulator diaphragm rupture and a serious accident. Replace and repair the defective regulator.

- Test the system for leaks at the cylinder-to-regulator fittings and all hose fittings with special leak detection solutions; bubbles indicate leaks.

55

Types of Flames

There are the three types of flames that different ratios of oxygen and acetylene can produce.

1. *Oxidizing flames* result when there is an excess of oxygen over acetylene. This flame will change the metallurgy of the weld pool metal by lowering the carbon content as it is converted to carbon dioxide.

2. *Neutral flames* result when there is just enough oxygen to burn all the acetylene present. This flame has the least effect on weld pool metal as only carbon monoxide and hydrogen combustion products result and is most frequently used in welding common materials.

3. *Carburizing flames* result when there is an excess of acetylene gas over the amount that can be burned by the oxygen present. The opposite of an oxidizing flame, it adds carbon to the weld pool and can change its metallurgy, usually adversely.

An oxidizing flame is significantly hotter than the other two flames, but is less useful as it will introduce more contaminants into the weld pool. An oxidizing flame is often used in braze welding or in fusion welding of heavy, thick parts with brass or bronze rod. In these applications, we are not concerned with weld pool contamination by carbon. An oxidizing flame is required for oxygen-fuel cutting.

Adjusting the Torch to a Neutral Flame

Open the acetylene valve no more than 1/16 turn and use a spark lighter to ignite the gas coming out of the tip. A smoky orange flame will be the result, Figure 3-28 (A).

Continue to open the acetylene valve until the flame stops smoking (releasing soot). Another way to judge the proper amount of acetylene is to open the acetylene valve until the flame jumps away from the torch tip, leaving about 1/16 inch gap (1.6 mm), Figure 3–28 (B). Then close the valve until flame touches the torch tip.

Open the oxygen valve slowly. As the oxygen is increased, the orange acetylene flame turns purple and a smaller, white inner cone will begin to form. With the further addition of oxygen, the inner cone goes from having ragged edges, Figure 3-28 (C), to sharp, clearly defined ones. The flame is now neutral and adding oxygen will make an oxidizing flame, Figure 3-28 (D).

If a larger flame is needed while keeping the same tip size, the acetylene may be increased and the oxygen further increased to keep the inner cone's edges sharp. This process of increasing the acetylene, then the oxygen is usually done in several cycles before the maximum flame available from a given tip is achieved. Adjusting the flame below the minimum flow rate for the tip orifice permits the flame to ignite inside the nozzle. This is *flashback* and makes a popping sound. If you need a smaller flame, use a smaller torch tip. See the section on flashback.

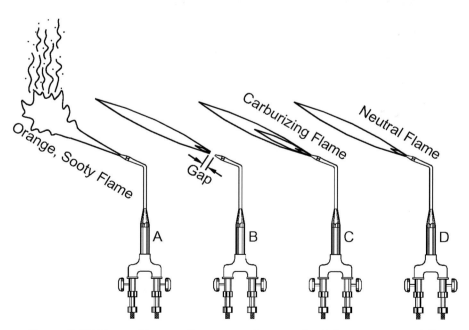

Figure 3-28 Shows flame adjustments from carburizing to a neutral flame

Flame Temperatures

The tip of the inner cone is the hottest part of the flame. The inner cone is where the optimum mixture of oxygen and acetylene burn. The outer envelope where any unburned acetylene burns with oxygen from the atmosphere. A neutral flame is when enough oxygen is present in the flame to be burning all of the acetylene gas and is used for most welding processes. See Figure 3-29.

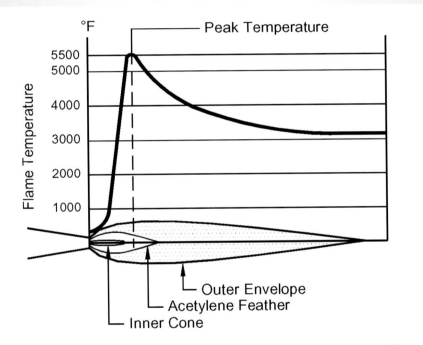

Figure 3-29 Graph of an oxyacetylene flame temperature profile

57

Lighting a Multi-Flame (Rosebud) Tip

This type of tip produces a large flame for heating metal prior to welding, bending, or brazing. When using a multi-flame tip you first set the acetylene pressure at or just below 15psi (1 bar) and the oxygen pressure at 30psi (2 bar); open the acetylene torch valve far enough to light the acetylene flame. Once the flame is ignited, open the acetylene valve until you have full flow of gas; now you can open the oxygen torch valve and adjust the flame to slightly carburizing. You may now use the multi-flame (rosebud) to heat materials but keep the sharp inner cone flame away from the material and only touch the carburizing flame to the material being heated. A heat sensing device such as a pyrometer or temperature sensing stick can be applied to the material to indicate the temperature of the material being heated.

Figure 3-30 A rosebud tip is often used to preheat materials prior to welding

Photo courtesy of Hobart Welders

How is the equipment shut down?

First turn off the oxygen and then the acetylene with the torch handle valves. Turning off the acetylene first can cause a flashback.

Turn off the oxygen and acetylene cylinder valves at the upstream side of the regulators.

Separately, open and reclose the oxygen and acetylene valves on the torch handle to bleed the remaining gas in the hoses and regulator into the atmosphere. Verify that both the high-pressure and low-pressure gauges on both regulators indicate zero.

Unscrew the regulator pressure adjustment screws on both cylinders in preparation for the next use of the equipment. The regulator screws should be loose but not about to fall from their threads.

What are the joint preparations for welding?

Refer to Figure 3-31.

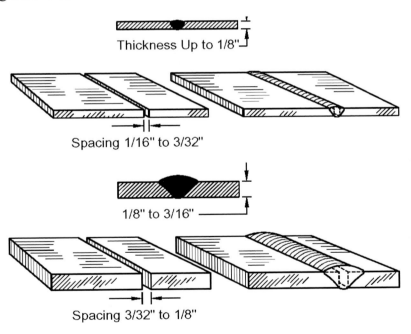

Figure 3-31 Preparation for OAW butt welds

How are welds judged?

For butt welds, here are examples of a correct weld, poor penetration weld, excessive reinforcement, undercutting, and excessive root reinforcement. See Figure 3-32.

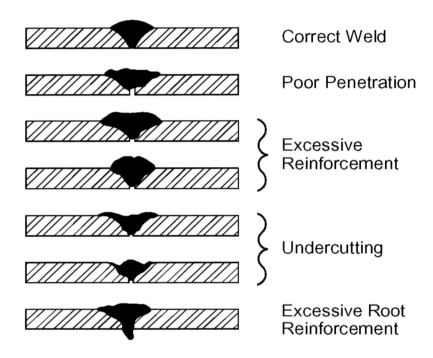

Figure 3-32 Correct and defective butt weld profiles

Producing a good weld bead is a combination of four main factors: the distance between the torch tip and the work, the angle at which you hold the torch, your speed when moving the torch along the weld area, and the heat produced by the torch. Getting everything right takes practice, so always test your technique on scrap metal first.

Figure 3-33 Create a puddle at the start of the bead. Keep the tip steady until a puddle begins to form. Begin making a circular motion with the torch, slowly moving the torch tip in the direction of the bead. Keep the distance to the work and the speed of your movements consistent.

61

Figure 3-34 Now try it with a filler rod. The goal is to intermittently dip the end of the rod into the puddle to add material to the weld. Dip the end of the rod into the puddle while making a circular motion with the torch. This will help blend the filler material into the weld. Withdraw the rod from the puddle, but keep it close to the end of the torch to keep it pre-heated. Don't directly heat the end of the rod with the torch.

Figure 3-35 The finished bead should be even throughout its length. The ripples created by the circular motion of the torch tip should be consistent.

Shielded Metal Arc Welding

Arc Welding and Stick Welding

This process is commonly known as "stick welding" or "arc welding," but it is officially labeled as "shielded metal arc welding" (SMAW) by the American Welding Society. In this process, an electric circuit is established between the welding power supply, the electrode, the welding arc, the work, the work connection, and back to the welding supply. Electrons flowing through the gap between the electrode and the work produce an arc that provides the heat to melt both the electrode metal and the base metal. Temperatures within the arc exceed 6,000°F (3,300°C). The arc heats both the electrode and the work beneath it. Tiny globules of metal form at the tip of the electrode and transfer to the molten weld pool on the work. As the electrode moves away from the molten pool, the molten mixture of electrode and base metals solidifies and the weld is complete.

How does the process work?

When can SMAW be used?

What is the basic equipment setup for SMAW?

What are the functions of SMAW electrodes
and how are they selected?

How is SMAW Equipment Set Up?

Stick welding tips from West Coast Customs

How does the process work?

courtesy of Lincoln Electric]

The electrode is coated with a flux. Heat from the electric current causes the flux's combustion and decomposition. This creates a gaseous shield to protect the electrode tip, the work, and the molten pool from atmospheric contamination. The flux contains materials that coat the molten steel droplets as they transfer to the weld and become slag after cooling. This slag also floats on the weld puddle's surface and solidifies over the weld bead when cool, where it protects the molten metal and slows the cooling rate. The flux coating on some electrodes contains metal powder to provide additional heat and filler to increase the deposition rate. The electrode flux and metal filler electrode determine the chemical, electrical, mechanical, and metallurgical properties of the weld as well as the electrode handling characteristics. Only 50% of the heat power furnished by the power supply heats the weld; the rest is lost to radiation, the surrounding base metal, and the weld plume.
See Figure 4-1.

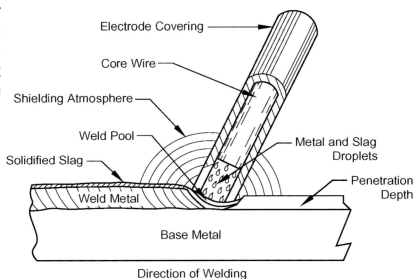

Figure 4-1.
The Shielded Metal
Arc Welding process

Electrode Covering

Core Wire

Shielding Atmosphere

Weld Pool

Solidified Slag

Weld Metal

Metal and Slag Droplets

Penetration Depth

Base Metal

Direction of Welding

When can stick welding be used?

Welds from 1/8 inch (3 mm) to unlimited thicknesses are possible. Thicknesses less than 1/8 inch (3 mm) can be joined but require much greater skill, while those over 3/4 inch (19 mm) are more economically done by other welding methods. Use this process for the following metals:

- Aluminum
- Bronze
- Carbon steel
- Cast iron
- Hard-facing

- High-strength steels
- Low-alloy steels
- Malleable iron
- Nickel
- Stainless steels

Little aluminum welding is done by SMAW as other, better processes exist. Note that some of these metals require preheat, post-heat, or both to prevent cracking.

Fiigure 4-2 Stick welding can handle a number of common maintenance tasks

Photo courtesy of Hobart

Pros and Cons of the Stick Welding Process

Advantages:

- Low-cost equipment.
- Can weld many different metals including the most commonly used metals and alloys.
- Relatively portable and useful in confined spaces.
- Same equipment welds thicknesses from 1/16 inch (16 gauge or 1.5 mm) to several feet in thickness with different current settings.
- Welds can be performed in any position.
- The process is less affected by wind and drafts than gas-shielded processes.
- There is no upper limit on thickness of metal to be welded.
- It can be performed under most weather conditions.

Disadvantages:

- Not suitable for metal sheets under 1/16 inch (1.5 mm) thickness.
- Operator duty cycle and overall deposition rate are usually lower with SMAW than with wire-fed processes because the SMAW process must be stopped when the electrode is consumed and needs to be changed.
- Not all of the electrode can be used; the remaining stub in the electrode handle must be discarded wasting one to two inches of electrode.
- Frequent stops and starts during electrode changes provide the opportunity for weld defects.

SMAW is not permitted in rain, snow, and blowing sand, and when the base metal is below 0°F (-18°C). Temporary shelters known as dog houses around the weld site permit welding during these conditions, increase the comfort of the welder, and help raise the weld steel temperature. Preheating the weld base metal to 70°F (21°C) with multi-flame (or rosebud) tips allows welding to proceed when the outside temperature is below 0°F (-18°C).

66

What is the basic equipment setup for SMAW?

The basic setup consists of
- Constant-current (CC) welding power supply
- Electrode holder, lead, and its terminals
- Ground clamp, lead, and its terminals
- Welding electrodes

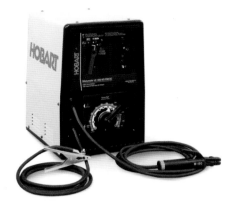

Welding Power Supplies

Constant-current (CC) welding power supplies are best for SMAW because these power supplies help maintain a constant current to the arc despite changing arc length during welding. This characteristic both gives the welder control over weld pool size and still limits maximum arc current. Some variation in arc length is inevitable as the welder moves the electrode along the weld while the CC welding supply insures a stable arc as these variations happen.

Figures 4-3 and 4-4
Two common SMAW setups that include the power supply, electrode holder, and ground clamps

Photos courtesy of Hobart Welders and Lincoln Electric

67

Types of current. The following types of current are used in stick welding.

- Alternating current (AC)
- Direct current electrode negative (DCEN = DCSP)
- Direct current electrode positive (DCEP = DCRP)

Direct current (DC) always provides the most stable arc and more even metal transfer than alternating current (AC). Once struck, the DC arc remains continuous. When welding with AC, the arc extinguishes and re-strikes 120 times a second as the current and voltage reverse direction. The DC arc has good wetting action of the molten weld metal and uniform weld bead size at low welding currents. For this reason it is excellent for welding thin sections. DC is preferred to AC on overhead and vertical welding jobs because of its shorter arc. Sometimes arc blow is a serious problem and the only solution may be to switch to AC. Most combination electrodes designed for AC or DC operation work better on DC.

DCEN stands for direct current electrode negative. This is also called *DCSP*, direct current straight polarity. DCEN produces less penetration than DCEP, direct current electrode positive, but has a higher electrode burn-off rate. See Figure 4-5.

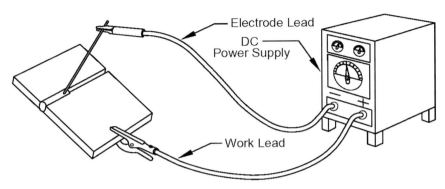

Electrode Lead
DC Power Supply
Work Lead

DC Electrode Negative (DCEN)
Replaces DC Straight Polarity (DCSP)

Figure 4-5 DCEN

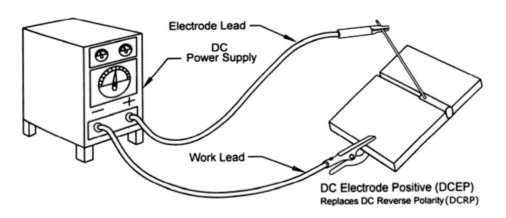

Electrode Lead
DC Power Supply
Work Lead

DC Electrode Positive (DCEP)
Replaces DC Reverse Polarity (DCRP)

Figure 4-6 DCEP

DCEP stands for DC-electrode positive. This is also called *DCRP*, direct current reverse polarity. DCEP is especially useful welding aluminum, beryllium-copper, and magnesium because of its surface cleaning action, which permits welding these metals without flux. DCEP produces better penetration than DCEN, but has a lower electrode burn-off rate. See Figure 4-6.

Ranking SMAW Welding Polarities. See Figure 4-7.
Going from highest to lowest heat into the weld:

DCEN
AC
DCEP

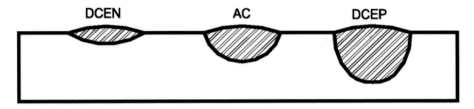

Figure 4-7 Effect of weld bead penetration at same welding current

Arc Blow

Whenever current flows, it creates a magnetic field around the conductor. Any shape of wire other than a straight line will produce an asymmetrical field along the wire—points where the field is either stronger or weaker than average. At currents above 600 amperes, the force created by the unevenness of the field may even cause the wire itself to move. More often, welders see the arc drawn in a particular direction like smoke in the wind. This is caused by residual magnetism in the part or the uneven magnetic field caused by current flowing through the part to the work connection. Some welders call this arc blow. See Figure 4-8.

Ways to reduce arc blow include:
- Using AC, instead of DC welding current.
- Moving the welding work lead to a position where you are welding away from the work lead connection.
- Using a shorter arc length.
- Clamping a steel block over the far end (unfinished end) of the weld.
- Welding away from the base metal edge, or toward a heavier tack or weld.
- Changing welding direction.

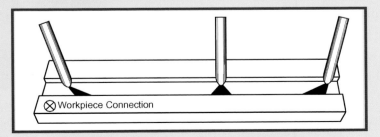

Figure 4-8 Arc blow on left and right arc.

69

Determining Current Settings of Welding Machines

Use the following steps:

1. Note the MEDIUM tap or coarse setting is in use with an output from 60 to 260 amperes.

2. This means that this tap will provide a minimum current of 60 amperes with the rotary knob set to zero. In the rest of the 60 to 260 ampere range an additional 200 amperes are gained by the ten steps on the knob.

3. Each of the ten knob points adds 20 amperes to the setting (200 amperes/ 10 knob steps = 20 amperes/each of the knob steps).

4. The knob is set to 6.5 steps, or 6.5 x 20 amperes plus the base of 60 amperes, or 190 amperes.

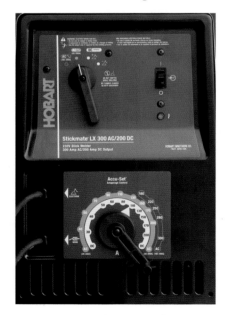

Figure 4-9 The amperage setting of a generic welding machine

70

Remember that these settings are likely to be approximate, and output current will vary with power line voltage variations, cable length, cable size, and other factors. The best output current setting may be slightly above or below the dial setting.

Electrodes and Amperage Range. Once an electrode class and diameter is selected for a task, use Table 4–1 to determine the manufacturer's recommended amperage range. Note that there will be some differences between recommended ranges for the same electrode class and diameter so it is important to consult the manufacturer's specification sheet for the electrode you are using. Also the material thickness will affect the current setting.

Usually it is best to use the lower current range setting. This is because the top of a current range is likely to have less current ripple (AC riding on the top of the DC) than the same current output setting from a higher current range, and will provide a smoother arc.

Figure 4-10 Note the AC and DC power settings on this stick welding machine
Photo courtesy of Hobart Welders.

Table 4-1 Recommended current rangers for various diameter electrodes

Lincoln Electric Product Name	AWS Class	Electrode Polarity	Sizes and Current Ranges (Amperes)					
			3/32"	1/8"	5/32"	3/16"	7/32"	1/4"
Fleetweld® 5P	E6010	DCEP	40-70	75-130	90-175	140-225	200-275	220-325
Fleetweld 35	E6011	AC	50-85	75-120	90-160	120-200	150-260	190-300
		DC±	40-75	70-110	80-145	110-180	135-235	170-270
Fleetweld 7	E6012	DC−	---	80-135	110-180	155-250	225-295	245-325
		AC	---	90-150	120-200	170-270	250-325	275-360
Fleetweld 37	E6013	AC	75-105	110-150	160-200	205-260	---	---
		DC±	70-95	100-135	145-180	190-235	---	---
Fleetweld 47	E7014	AC	80-100	110-160	150-225	200-280	260-340	280-425
		DCEN	75-95	100-145	135-200	185-235	235-305	260-380
Jetweld® LH-70	E7018	DC+	70-100	90-150	120-190	170-280	210-330	290-430
		AC	80-120	110-170	135-225	200-300	260-380	325-530
Jetweld 3	E7024	AC	---	115-175	180-240	240-315	300-380	350-450
		DC±	---	100-160	160-215	215-285	270-340	315-405
Jetweld 2	E6027	AC	---	---	190-240	250-300	300-380	350-450
		DC±	---	---	175-215	230-270	270-340	315-405
Jetweld LH-3800	E7028	AC	---	---	180-270	240-330	275-410	360-520
		DC+	---	---	170-240	210-300	260-380	---

Welding Power Supply Maintenance

The power supply should be blown out with compressed air to remove slag, grinding dust, and other dust from its interior. If you use your equipment frequently, perform the maintenance on a monthly basis.

Electrode Holders and Work-Lead Connections

Besides holding the electrode in place, the electrode holder insulates the welder from the welding power supply voltage; thermally insulates the welder from the conducted heat of the electrode; and makes a secure electrical connection between the welding cable and the electrode with a minimum of voltage drop. Poor gripping of the electrode by the electrode holder's jaws will cause poor conduction of electricity, and power intended for the weld will be lost in the holder. The holder will also heat up and prevent the operator from holding it. Eventually the heat will deteriorate its electrical insulation properties exposing the welder to possible electric shock. Most electrode holders have notches where the electrode conductor end fits and will not move. When selecting, it is best to pick the smallest size that can be used without overheating as it will be the lightest weight and least tiring to work with.

Figure 4-11 Position the bare metal end of the electrode in the holder. Many stick welding electrode holders can clamp the electrode into a variety of positions

Judging Arc Length

Although the correct arc length is usually 1/8 inch, one should never exceed the diameter of the electrode. The sound of the arc is an excellent indicator of arc length. The proper arc length sounds like the crackling of bacon and eggs frying. Too short an arc, makes a sputtering sound; too long an arc, a humming sound.

72

What are the functions of stick welding electrodes and how are they selected?

Stick welding electrodes contain a solid or cast metal core wire covered by a thick flux coating. Every change of flux composition and thickness alters the operating characteristics. These changes plus current type and polarity determine how the rod will handle and what type bead it will deposit. They are made in lengths from 9 to 18 inches (230 to 460 mm). Some electrodes are made with a metallic tube containing a mixture of metal powders.

Functions of the Rod Coating

The principal function of the coating is to provide a gas stream to shield the molten weld pool from atmospheric oxygen, hydrogen, and nitrogen contamination until it solidifies. But the coating also

• Supplies scavengers, deoxidizers, and fluxing agents to clean the weld and prevent excessive grain growth in the weld material.

• Provides chemicals to the arc that control the electrical characteristics of the electrode: Current type, polarity, and current level.

73

• Covers the finished weld with slag, a protective covering to control weld cooling rate, protect the cooling materials of the weld from the atmosphere, and control bead shape.

• Adds alloying materials to the weld pool to enhance weld properties.

Stick Welding Electrode Classification

Both the American National Standards Institute (ANSI) in collaboration with the American Welding Society (AWS) and ASME International classify electrodes. Both provide convenient and effective ways of choosing welding electrodes for carbon steel. But it is important to check the electrode manufacturer's specifications as well.

The ANSI/AWS System. Electrodes for carbon and low-alloy steel have the ANSI/AWS classification number stamped directly on the flux of the electrode in one or more places. Table 4-2 explains the coding and 4-3 describes the electrodes in each classification. Color coding SMAW *coated* electrodes is obsolete, but un-coated electrodes used for surfacing are still color coded. Some manufacturers use color codes to identify their electrodes and are explained in their data sheets. ANSI/AWS classifies electrodes used for welding other metals but under different specifications. Again, when in doubt, check the manufacturer's data sheets.

Figure 4-12 and 4-13 Electrodes come in sealed containers

Photos courtesy of Lincoln Electric and Hobart Welders.

The tables on the following two pages describe the electrode classifications systems of AWS. Table 4-2 explains the coding found on electrodes. Table 4-3 describes characteristics of each electrode classification.

Table 4-2 The coding used in AWS electrode classification system for carbon and low alloy steel electrodes

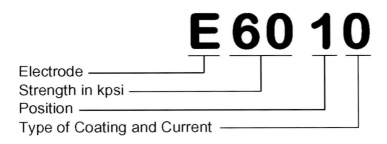

E 60 10

Electrode
Strength in kpsi
Position
Type of Coating and Current

E 8018-B1H4R

Electrode
80,000 PSI Min.
All Position
For AC or DCEP
Chemical Composition of Weld Metal Deposit

Diffusible Hydrogen Designator Indicates the Maximum Diffusible Hydrogen Level Obtained with the Product.

Moisture Resistant Designator Indicates the Electrode's Ability to Meet Specific Low **Moisture Pickup Limits** under Controlled Humidification Tests.

75

Position

1. Flat, Horizontal, Vertical, Overhead
2. Flat and Horizontal Only
3. Number 3 Position Is Not Designated
4. Flat, Horizontal, Vertical Down, Overhead

Types of Coating and Current

Digit	Type of Coating	Welding Current
0	Cellulose Sodium	DCEP
1	Cellulose Potassium	**AC or DCEP or DCEN**
2	Titania Sodium	AC or DCEN
3	Titania Potassium	AC or DCEP
4	Iron Powder Titania	**AC or DCEN or DCEP**
5	Low Hydrogen Sodium	DCEP
6	**Low Hydrogen Potassium**	AC or DCEP
7	Iron Powder Iron Oxide	**AC or DCEP or DCEN**
8	**Iron Powder Low Hydrogen**	AC or DCEP

DCEP - Direct Current Electrode Positive
DCEN - Direct Current Electrode Negative

Chemical Composition of Weld Deposit

Suffix	%Mn	%Ni	%Cr	%Mo	%V
A1				1/2	
B1			1/2	1/2	
B2			1-1/4	1/2	
B3			2-1/4	1	
C1		2-1/2			
C2		3-1/4			
C3		1	.15	.35	
D1& D2	1.25-2.00			.25-.45	
G[1]		.50	.30 min	.20 min	.10 min

(1) Only one of the listed elements required.

Table 4-3 The characteristics of AWS electrode classifications

AWS Class	ASME System	Current and Polarity	Welding Positions	Type of Covering	Type of Arc	Degree of Penetration	Surface Appearance	Type of Slag	Character of Slag
EXX10	F-3	DC, positive electrode	All	High-cellulose, sodium	Digging	Deep	Flat, wavy	Organic	Thin
EXX11	F-3	AC or DC, electrode positive	All	High-cellulose, potassium	Digging	Deep	Flat, wavy	Organic	Thin
EXX12	F-2	Ac or DC, electrode positive	All	High-titania, sodium	Medium	Medium	Convex, rippled	Rutile	Thick
EXX13	F-2	AC or DC, either Polarity	All	High-titania, potassium	Soft	Shallow	Flat or concave, smooth ripple	Rutile	Medium
EXX14	F-2	DC, either polarity or AC	All	Iron-powder, titania	Soft	Medium	Flat, slightly convex, smooth ripple	Rutile	Easily removed
EXX15	F-4	DC, electrode positive	All	Low-hydrogen, sodium	Medium	Medium	Flat, wavy	Low-hydrogen	Medium
EXX16	F-4	AC or DC, electrode positive	All	Low-hydrogen, potassium	Medium	Medium	Flat, wavy	Low-hydrogen	Medium
EXX18	F-4	DC, electrode positive or AC	All	Low-hydrogen, potassium, iron powder	Medium	Shallow	Flat, smooth, fine ripple	Low-hydrogen	Medium
EXX20	F-1	DCEN or AC for horizontal fillets; DC, either polarity, or AC for flat work	Horizontal fillets & flat	High-iron oxide	Digging	Medium	Flat or concave, smooth	Mineral	Thick
EXX22	F-1	DC, either polarity or AC	Horizontal, flat	High-iron oxide	Soft, smooth	Medium	Flat or slightly convex	Mineral	Medium
EXX24	F-1	AC or DC, either polarity	Horizontal fillets and flat	Iron powder, titania	Soft	Shallow	Slightly convex, very smooth, fine ripple	Rutile	Thick
EXX27	F-1	AC or DC, electrode negative	Horizontal fillets and flat	High-iron oxide, iron powder	Soft	Medium	Flat to slightly concave, smooth fine ripple	Mineral	Thick
EXX28	F-1	AC or DC, electrode positive	Horizontal fillets and flat	Low-hydrogen, potassium, iron powder	Medium	Shallow	Flat, smooth, fine ripple	Low-hydrogen	Medium
EXX48	F-4	Ac or DC, electrode positive	Flat, horizontal, vertical-down, overhead	Low-hydrogen, potassium, iron powder	Soft	Shallow	Concave. Smooth	Low-hydrogen	Thin

The ASME System. ASME classifies coated electrodes for carbon and mild steel work in four groups.

F-1, High Deposition Group (also called *Fast–Fill*)
F-2, Mild Penetration Group (also called *Fill–Freeze*)
F-3, Deep Penetration Group (also called *Fast–Freeze*)
F-4, Low Hydrogen Group

F-1. High Deposition Group

- The electrode coating contains 50% iron powder by weight, so these electrodes. produce a higher weld deposition per electrode than any members of the other groups.
- Dense slag and slow cooling make this group useful only for flat and horizontal fillet.
- Produces smooth ripple-free bead with little spatter.
- Heavy slag produced is easily removed.

F-2 Mild Penetration Group

- Electrodes have a titania, rutile, or lime-based coating.
- They are excellent for welding sheet steel under 3⁄16 inch (5 mm) thick where high speed travel with minimum skips, slag entrapment, and undercut are required. These are often used with DCEN.

F-3 Deep Penetration

- Electrodes have a high cellulose coating that produces deep penetration and a forceful arc. They may also contain iron powder, rutile, and potassium.
- The weld solidifies rapidly for use in all positions.
- Excellent for welding mild steel in fabrication and maintenance work.
- Best choice on dirty, painted, or greasy metal.
- Light slag.
- Especially good for vertical-up, vertical-down, and overhead and open root welding.

F-4 Low Hydrogen Group

- Resistance to hydrogen inclusions and underbead cracking in medium to high carbon steels, hot cracking in phosphorus-bearing steels, and porosity in sulfur-bearing steels.
- Excellent for X-ray quality welds and mechanical properties.
- Less preheat than other electrodes.
- Reduced likelihood of underbead and micro-cracking of high carbon and low alloy steels and on thick weldments.
- Can produce excellent multiple pass, vertical, and overhead welds in carbon steel plate.
- Best choice for galvanized metal.

Common Dimensions of SMAW Electrodes

The most common length is 18 inches (460 mm). The common diameters (of the core wire) are in inches (mm):

1/16 (1.6 mm)	3/32 (2.3 mm)	1/8 (3.2 mm)	5/32 (3.9 mm)
3/16 (4.7 mm)	7/32 (5.6 mm)	1/4 (6.3 mm)	5/16 (7.9 mm)

Selecting Electrodes

This is not a simple decision since there are trade-offs between speed, total welding cost, and weld strength. Electrode selection is a matter of matching the operating characteristics of the electrode to the job requirements as well as to the possible need for low hydrogen electrodes. In general, the electrode rod steel should have the same or higher tensile strength as the base metal and similar chemical properties. Here are the major factors in the selection:

- Welder's skill
- Properties of the base metal
- Position of the weld joint
- Type of joint
- Type of power supply
- Tightness of the joint fit-up
- Total amount of welding needed

Value of including iron powder in the electrode coating

- Welding heat is used to melt the core and coating, not excess areas of the base metal.
- Iron in the flux coating adds to the weld deposit and increases deposition rate.
- Drag technique of welding is used.

Problems with Moisture and Some Electrodes

Water absorbed out of the atmosphere by the flux on low hydrogen electrodes will introduce hydrogen into the weld, causing cracking and brittleness. Dry electrodes can take from 30 minutes to four hours to pick up enough water (and the water's hydrogen) to affect weld quality. For this reason, professional welding shops keep low-hydrogen electrodes in drying ovens after removing them from sealed electrode dispensers until they are used. This option is usually not available to the home welder. If using this type of electrode, check the manufacturer's data sheets for instructions on handling the electrodes.

79

Determining Electrode Size

The variables to consider include

- Joint thickness
- Welding position
- Type of joint
- Welder's skill

The best choice of electrode will be the one that will produce the weld required in the least time. Usually larger diameter electrodes are used for thicker materials and flat welds where their high deposition rates can be an advantage. In non-horizontal positions, smaller electrodes are used to reduce the weld pool size as gravitational forces are a factor working against the welder's skill.

Selecting an Electrode for Carbon Steel

Determine which of the Figs.4-14 through 4-18, best illustrates the joint, position, and metal thickness of the task you have.

Read the electrode recommendation above the joint.

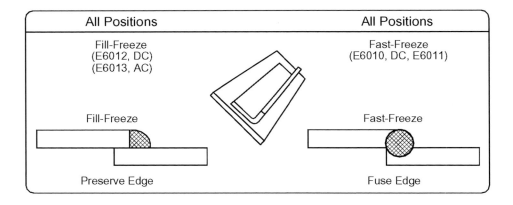

Figure 4-14 Sheet metal joints

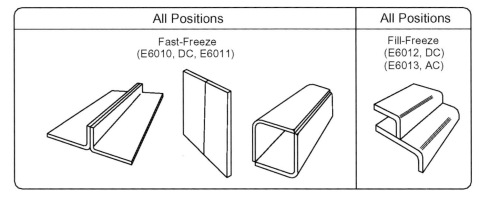

Figure 4-15 Sheet metal joints

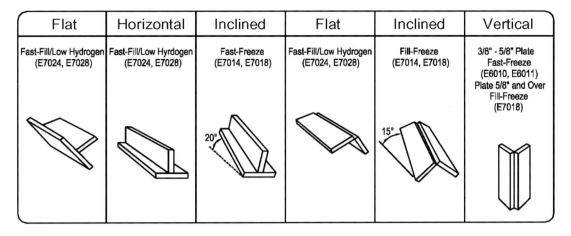

Flat	Horizontal	Inclined	Flat	Inclined	Vertical
Fast-Fill/Low Hydrogen (E7024, E7028)	Fast-Fill/Low Hyrdogen (E7024, E7028)	Fast-Freeze (E7014, E7018)	Fast-Fill/Low Hydrogen (E7024, E7028)	Fill-Freeze (E7014, E7018)	3/8" - 5/8" Plate Fast-Freeze (E6010, E6011) Plate 5/8" and Over Fill-Freeze (E7018)

Figure 4-16 Light plate joints

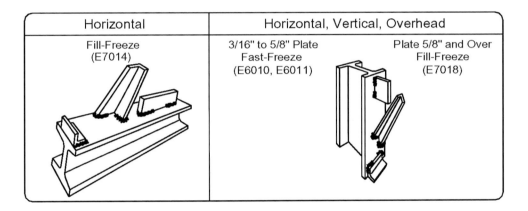

Horizontal	Horizontal, Vertical, Overhead	
Fill-Freeze (E7014)	3/16" to 5/8" Plate Fast-Freeze (E6010, E6011)	Plate 5/8" and Over Fill-Freeze (E7018)

Figure 4-17 Short fillets under 6 inches (150 mm) in length and having a change of direction on 3/16 inch (4.7 mm) or thicker plate

Horizontal, Vertical, Overhead	Flat	Flat	Flat
3/16" to 5/8" Plate Fast-Freeze (E6010, E6011) Plate Over 5/8" Fill-Freeze (E7018)	3/8" and Thicker Fast-Fill-Low Hydrogen (E6027, E7028)	3/8" and Thicker Root Pass Fill-Freeze (E7018) All Other Passes Fast-Fill-Low Hydrogen (E6027, E7028)	3/8" and Thicker Root Pass Fill-Freeze (E7018) All Other Passes Fast-Fill-Low Hydrogen (E6027, E7028)

Figure 4-18 Heavy plate welds

81

How is Stick Welding Equipment Set Up?

• Locate the welding power supply near the work, and note the location of the AC power shut-off switch in case of emergency. Make sure the welding power supply is grounded. If an engine-generator is the welding power source, locate the engine shut-off switch.

• Make sure the areas around the power supply and the work are dry.

• Remove flammable materials near the spark stream area and insure fire fighting materials are close at hand.

• Based on the job requirements or drawings, select the electrode material and diameter.

• Set welding polarity and welding current on welding machine.

• Stretch out the welding cables, and attach the work lead to the work securely; clean the grounding area if necessary.

• Set out the electrodes, welding safety equipment (helmet, cap, gloves, leathers); you should have had safety glasses on since the first step and they should remain on under your welding helmet.

• Turn on the welding machine, insert the electrode in the holder, drop your welding helmet, strike the arc, and begin welding.

Striking the Arc

Scratch-start, usually used for AC, is performed by tilting the electrode about 15° in the direction of travel and while drawing the electrode above the work as if striking a match, allowing it to momentarily touch the work; this strikes the arc. The electrode is withdrawn above the work to a height equal to its own diameter.

Tapping-start, usually used for DC, is made by lowering the electrode quickly until an arc forms and then raising the electrode its own height above the work.

Re-starting an Arc on an Existing Weld. Strike an arc about an inch further along the weld path than the interrupted weld bead. Then carry the arc back to the previous stopping point and begin welding again. The new weld bead will cover and eliminate the marks of the strike.

Determining Work and Travel Angle. The joint type and welding position determine these variables. See Table 4–5.

Filter Plate Shade Number for SMAW

The shade of the filter lens depends on the arc current and the electrode diameter; use Table 4–4.

Figure 4-19 Be sure your helmet is outfitted with the correct filter plate when stick welding

Table 4.4 Filter Plate Shade Number Based on Electrode Diamter

Welding Electrode Diamter (inches)	Welding Electrode Diameter (mm)	Filter Plate Shade Number
Up to 5/32	4	10
3/16-1/4	4.8-6.4	12
Over 1/4	6.4	14

Table 4.5 SMAW Orientation and Welding Technique for Carbon Steel Electrodes

Type of Joint	Welding Position	Working Angle (degrees)	Travel Angle (degrees)	Technique of Welding
Groove	Flat	90	5-10*	Backhand
Groove	Horizontal	80-100	5-10	Backhand
Groove	Vertical-Up	90	5-10	Forehand
Groove	Overhead	90	5-10	Backhand
Fillet	Horizontal	45	5-10*	Backhand
Fillet	Vertical-Up	35-45	5-10	Forehand
Fillet	Overhead	30-45	5-10	Backhand

*Travel angle may be 10-30° for electrodes with heavy iron castings.

Stick welding is often referred to as arc welding, which can be confusing for the novice because all modern welding techniques employ an electric arc to produce the heat necessary to fuse metal. Stick, or Shielded Metal Arc Welding, was one of the first and the name has stuck with the process.

Figure 4-20 To start welding, set the welding polarity and current on the welding machine.
Consult the electrode manufacture's data sheet.
Tables 4-1 and 4-3 in this chapter also provide information on machine settings

Figure 4-21 Clamp the electrode into position. Keep extra electrodes handy because they are consumed quickly. You won't be able to weld much more than about 8 to 10 inches, depending on the type of joint, with each electrode

Figure 4-22 When starting a bead, circle the rod a few times at your start point. This heats up the metal and allows a weld pool to form. You can either push or pull the electrode when welding. Experiment with both. When welding, maintain a constant arc length, which is the distance between the electrode and the work. The length should be slightly smaller than the diameter of the electrode you are using

Figure 4-23 What you see on this tack weld is slag that covers the weld. The slag protects the weld from impurities while it is fusing the metal together. Remove the slag with a chipping hammer and follow with a wire brush if necessary. When restarting a weld, chip away some of the slag at the end of the previous section. Then restart a little ahead of the finished area and work back to tie the two sections together

Wire Feed Welding Processes

MIG Welding and Flux Cored Arc Welding

Most commonly known as MIG, for metal inert gas, or wire feed welding, its official designation is Gas Metal Arc Welding (GMAW). In this process, the welding equipment continuously feeds a wire electrode to the welding gun. As the electrode is consumed during the course of the weld, new wire takes its place. A shielding gas, also dispensed through the welding gun, keeps the weld free of impurities that could weaken it. Flux Cored Arc Welding (FCAW) is a similar process, but rather than use a shielding gas, the electrodes consist of a thin-walled tube filled with flux, which protects the weld. Each process will be discussed in this chapter.

How does the MIG process work?

What metals can MIG weld?

What is the basic equipment for MIG Welding?

How do you select MIG electrodes?

What part does shielding gas play in the MIG process?

How is the welding wire transferred to the welding pool?

How is MIG equipment set up?

What are some joint preparation steps for MIG Welding?

How does the orientation of the electrode during welding affect the weld?

What are typical MIG Welding problems and solutions?

How does the flux cored arc welding process work?

What equipment is needed for an FCAW welding outfit?

How do you select FCAW electrodes?

What are typical FCAW problems and solutions?

How is FCAW equipment set up?

MIG welding tips from West Coast Customs

How does the MIG process work?

The welder positions the electrode wire coming from the center of the contact tip to where the weld is to begin, actually touching the welding wire to the base metal. When the welder drops his hood and squeezes the trigger on the welding gun, three events happen simultaneously:

1. The trigger turns on the welding power supply; welding cable conductors feeding the gun apply this voltage to the copper contact tip within the gun, and then to the electrode wire, striking the arc.
2. The wire feed mechanism begins feeding the welding wire from the spool through the welding gun cable and out the contact tip into the weld pool. Note that the welding gun cable is really a bundle of cables: power cables, control cables, electrode wire liner, shielding gas line, and possibly, cooling water and smoke extractor lines.
3. An electrically operated valve (solenoid) opens and feeds shielding gas from the regulator/flowmeter to surround the electrode and weld pool, shielding them from the atmosphere, particularly oxygen and nitrogen. See Figure 5-1.

Welding begins as the section of electrode wire between the tip and the base metal is heated and deposited into the weld. As the wire is consumed, the feed mechanism supplies more electrode wire at the pre-adjusted rate to maintain a steady arc. The welder manipulates the gun and lays down the weld in the desired pattern.

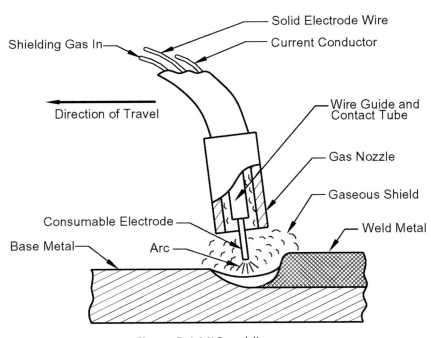

Figure 5-1 MIG welding gun

To stop the process when the weld is completed, the welder releases the trigger shutting off the welding current, wire feed, and shielding gas.

What metals can MIG weld?

- Aluminum
- Carbon steel
- Copper
- Low alloy steels
- Magnesium
- Nickel
- Stainless steels
- Titanium

Pros and Cons of the MIG Process

Advantages

- MIG welding is the only consumable process that can weld most commercial alloys.
- MIG welding, with its continuous wire electrode, overcomes the start-and-stop cycle of SMAW and leads to fewer discontinuities and higher deposition rates.
- Since there is no need to stop welding to change consumed electrodes, long continous welds can be done manually.
- All welding positions can be used using the short-circuit transfer mode.
- Significantly higher utilization of filler metal than SMAW (stickwelding): There is no end loss as with the upper, unconsumed end of an SMAW electrode.
- With spray transfer, deeper penetration with higher deposition than SMAW and may permit smaller fillets of the same strength.
- Metals as thin as 24 gauge (0.023 inch or 0.5842 mm) may be welded.
- Easier to learn than most other welding processes.
- There are no practical thickness limitations.
- There is very little spatter and no slag with properly adjusted equipment. Many manufacturers paint or plate over MIG welds with little or no additional surface preparation.

Disadvantages

- Welding equipment can be more expensive, more complicated, and slightly less portable than SMAW.
- MIG welding is more difficult to use in tight quarters as the gun is large and the gun cable is somewhat stiff and inflexible.
- The large size of the MIG gun combined with the 1/2 to 1 inch (12.7 to 25 mm) stickout of the electrode wire makes it harder to see the arc and achieve quality welds.
- MIG welding outdoor use is limited to very calm days or where shielding screens can be used to prevent the shielding gas from being blown away. MIG cannot be performed outdoors in greater than a 5 mile/hour (8 km/hour) breeze.

What is the basic equipment for MIG welding?

The basic equipment consists of:

• Constant-voltage (CV) welding power supply

• Wire feeder containing wire feed motor, spool support, wire feed drive rolls, and associated electronics (may be an integral part of welding power supply or separate unit)

• Welding gun and its cable

• Work lead clamp, work lead (cable), and its terminals

• Welding wire

• Flow regulator or flow meter for shielding gas(es)

• Compressed gas cylinder

See Figures 5-2 and 5-3.

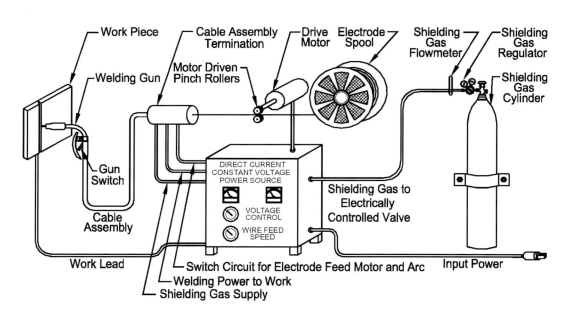

Figure 5-2 MIG welding outfit

Figure 5-3 Some welding machines can handle a number of different types of welding

Photo courtesy of Miller Electric.

Additional Equipment

Sometimes setups include:
- Water cooler
- Smoke evacuator

(MIG) and FCAW welding guns operated continuously at high amperage levels get so hot they literally melt the gun from the heat of the weld and the heat from the contact tip. This situation arises during spray-metal transfer, which is discussed later. The solution is to feed cooling water into the nozzle to keep its temperature down. The hot water from the tip is then recirculated back to the water cooler and used again. The water cooler consists of a pump, fan, and internal spray, much like an evaporative cooler. Alternatively, city water may be used to cool the welding gun and then discharged into a drain; this approach eliminates the need for a water cooler.

Some GMAW and FCAW guns have built-in systems for capturing and removing welding smoke at the nozzle of the welding gun. The smoke reduction adds to welder safety, comfort, and visibility, but the evacuator also adds weight to the gun and stiffness to the welding cable.

Welding Power Supplies

DCEP (direct current electrode positive) is always used. This is because the three metal transfer modes that actually move metal ions through the arc (short-circuit, globular, and spray transfer modes—more on those later), move positively charged metal ions. These positive ions must travel through a positive to negative voltage field—just what DCEP provides.

Arc Voltage. The best way to set arc voltage is to begin at the recommended voltage based on welding tables from AWS or the electrode manufacturer's data sheets. The base metal, type of metal transfer, and shielding gas will yield a starting point and exact refinement will come from several trial runs.

The goal is to control arc length, but it is hard to measure and closely related to arc voltage, so we choose to control arc voltage. Most welding specifications indicate arc voltage setting, not arc length. This is because with all other variables held constant, arc length is proportional to arc voltage. Arc voltage depends on materials welded, shielding gas, and metal transfer mode.

Amperage. Welding current (amperage) is determined by the wire feed rate, which is the speed the electrode emerges from the welding gun and is usually marked on the speed control or digital readout. The higher the electrode wire feed rate, the more current the constant voltage welding power supply must provide. The welding current on a constant voltage power supply cannot be directly adjusted. With all other welding variables held constant, an increase in wire feed speed (and thus welding current) will:

92 A

- Increase depth and width of weld penetration

- Increase deposition rate

- Increase weld bead size

How do you select MIG electrodes?

MIG electrodes are usually solid, bare wire similar in composition to the base metal being welded. Some electrode wire for welding ferrous metals has a thin copper plating to facilitate the drawing of the wire when made at the factory. Some wire makers would have you believe this coating is added to prevent rust and to make the wire run better; it is not.

Silicon is also added to maintain metal integrity at high arc temperatures. As an aggressive scavenger, it combines with unwanted elements and forms a glaze on the weld surface. It pops off when cool and is another reason to wear safety glasses at all times.

Common Dimensions of MIG Electrode Wires

The common diameters in inches:

0.020	0.062 (or 1/16)	0.313 (or 5/16)
0.025	0.094 (or 3/32)	0.375 (or 3/8)
0.030	0.125 (or 1/8)	0.500 (or 1/2)
0.035	0.188 (or 3/16)	
0.045	0.250 (or 1/4)	

Wire Feed Rate and Weld Deposition Rate

The wire feed rate, measured in inches/sec, or pounds/hour, is the speed at which the electrode wire emerges from the welding gun. The deposition rate is the weight of welding wire going into the weld. It is nearly always less than the wire feed rate. The difference between the two is slag, spatter, or fumes from the electrode wire. The deposition efficiency is the ratio of the weld deposition rate to the wire feed rate. GMAW with an argon shielding gas can reach 98% deposition efficiency.

Figure 5-4. This is a typical example of a spool of MIG welding wire

Photo courtesy of Hobart Welders.

Identifying MIG Welding Wire

The American Welding Society's GMAW electrode identification system is shown in Table 5-1 below.

Table 5-1 The AWS designation for GMAW welding wire

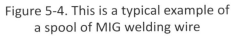

Strength   **Chemical Composition**

E R XX S - X

Electrode Rod   Solid Wire

GMAW Electrode Identification System

93

"ER" designates the wire as being both an electrode and a rod, meaning it may conduct electricity (electrode) or simply be applied as a filler metal (rod) when used with the GMAW process.

The characteristics of the most common GMAW electrodes for carbon steel meeting AWS/ASTM Specifications are shown in Table 5-2.

Table 5-2 MIG Electrode Wires for Mild Steel		
AWS Designation	**Characteristics**	**Shielding Gas(es)**
ER70S-2	Contains deoxidizers that permit welding on thin rust coatings. Can weld in any position. Excellent for out-of-position short-circuit welding. Makes excellent welds in all mild steels.	Ar-O_2 Ar-CO_2, CO_2
ER70S-3	Single or mulitpass beads. Base metal must be clean. Preferred for galvanized metals. Used on autos, farm equipment, and home appliances. Has better wetting action and flatter beads than E70S-2.	Ar-CO_2, CO_2
ER70S-4	Used for structural steels like A7 and A36. Used for both short-circuit and spray. Used in ship building, pipe welding, and pressure vessels. Flatter and wider beads that E70S-3.	Ar-O_2 Ar-CO_2, CO_2
ER70S-5	Used on rusty steel. Not recommended for short-circuit transfer mode. Flat position only.	Ar-O_2 Ar-CO_2, CO_2
ER70S-6	General purpose wire used on sheet metal. Will weld through thin rust. Welds all positions. Used with high welding currents.	Ar-O_2 Ar-CO_2, CO_2
ER70S-7	Used on heavy equipment and farm implements. Welds in all positions.	Ar-CO_2, CO_2

Ar = Argon CO_2 = Carbon Dioxide

Electrode Wire Handling

The three most common methods of moving electrode wire from the spool to the welding area.

- Drive wheels in the wire feeder—one or more pairs of drive wheels in the wire feed machine grip the electrode wire and push it through the wire feed tube to the welding gun. This is by far the most common arrangement and works well except for very soft electrodes like aluminum.

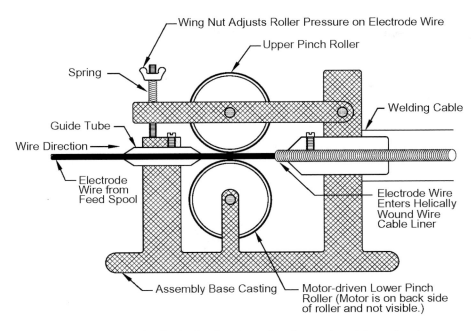

Wing Nut Adjusts Roller Pressure on Electrode Wire

Upper Pinch Roller

Spring

Welding Cable

Guide Tube

Wire Direction →

Electrode Wire from Feed Spool

Electrode Wire Enters Helically Wound Wire Cable Liner

Assembly Base Casting

Motor-driven Lower Pinch Roller (Motor is on back side of roller and not visible.)

Figure 5-5 Typical wire feed mechanism using drive wheels

95

• Spool gun/torch—the torch assembly contains both a small spool of electrode wire and a DC motor to drive it out the contact tube. These work well for handling soft wires like aluminum, but the penalty is the increased weight the welder must manipulate. See Figure 5-6.

Figure 5-6 Spool guns work well for handling soft wire

Photo courtesy of Miller Electric.

• Pull gun—the torch assembly contains a small DC motor and drive wheels, which assist in pulling the soft wire through the cable liner and out the torch tip. This has the advantage of reduced torch weight with the ability to handle soft electrodes. See Figure 5-7.

Figure 5-7 Pull gun

Supplying Power to the Electrode

Welding current travels down the MIG cable to the welding gun assembly via a copper cable and then transfers to a copper transfer tube, which puts it onto the electrode. See Figure 5-8.

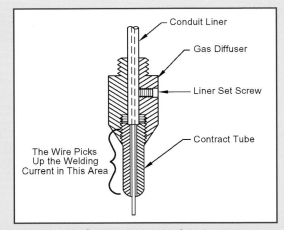

Figure 5-8 Detail of GMAW gun tip showing copper transfer tube, which feeds welding current into the electrode wire

What part does shielding gas play in the MIG process?

The main function of shielding gas is to prevent the atmospheric oxygen and nitrogen from reaching the molten metal of the weld. Most metals in the molten state combine with atmospheric oxygen and nitrogen to form oxides and nitrides. These two classes of compounds can leave the weld with porosity, embrittlement, or trapped slag and greatly reduce its mechanical properties.

Shielding gases may also:
- Stabilize the arc
- Control the mode of metal transfer
- Establish the penetration level and weld bead shape
- Enhance welding speed
- Minimize undercutting
- Clean oil or mill scale just prior to welding
- Control weld metal mechanical properties

Figure 5-9 Here is a portable MIG setup with shielding gas tank

Shielding Gases Used for MIG Welding

The following inert gases will not react with other chemical elements.

Argon (Ar)—often used for out-of-position welding because of its lower heat conductivity; it is ten times heavier than helium so it blankets the weld better than helium.

Helium (He)—this gas is an excellent heat conductor bringing heat from the arc into the weld area. It is used where high heat input is needed as in joining thick sections and when welding copper and aluminum, which are excellent conductors of heat and remove heat rapidly from the welding zone.

The following reactive gases will form compounds with other chemical elements and are used to achieve specific objectives at low concentrations in GMAW.

Carbon dioxide (CO_2)—causes better metal transfer, lower spatter, more stable arc, and improved flow of metal to reduce undercutting.

Oxygen (O_2)—blended with argon in small percentages to reduce volts.

Hydrogen (H_2)—in small percentage concentrations removes light rust and eliminates surface cleaning.

Nitrogen (N_2)—used on copper.

97

Gas Mixtures Commonly Used for MIG Welding

Argon + Helium
Argon + Oxygen
Argon + Carbon dioxide
Helium + Argon + Carbon dioxide

Mixtures produce better arc stability, lower spatter, and better bead structure than any one gas used by itself. By mixing gases we can get the best performance properties each component gas can produce.

See Figure 5-10 for the characteristics produced by the various shielding gases/combinations.

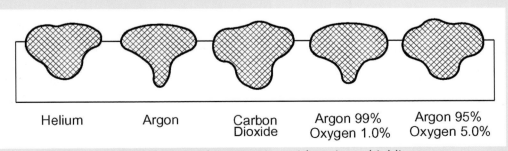

| Helium | Argon | Carbon Dioxide | Argon 99% Oxygen 1.0% | Argon 95% Oxygen 5.0% |

Figure 5-10 Bead shape and penetration with various shielding gases

Shielding Gas Metering Systems

There are two types of systems. They are pressure regulator with high- and low-pressure gauges and pressure regulator with ball-in-tube flow meter. See Figure 5-11.

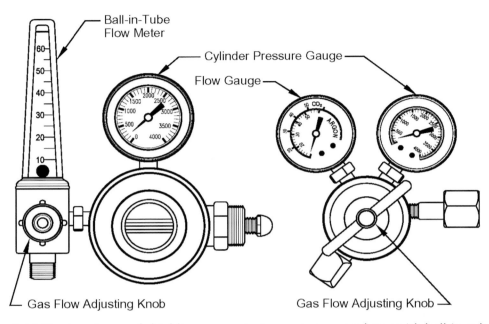

Figure 5-11 The two types of shielding-gas metering systems: regulator with ball-in-tube flow meter (left) and regulator with gauges (right) both control flow in cubic feet per hour
These are not pressure gauges.

Figure 5-12 The shielding gas prevents impurities in the atmosphere from reaching the molten metal of the weld

Photo courtesy of Lincoln Electric.

Adjusting Shielding Gas

Adjust gas flow according to either AWS specification or manufacturer's data sheet. If neither is available, begin at about 20-25 ft^3/hour (11.8 l/min); increase gas flow until visible signs of weld porosity cease.

Getting an effective inert gas shield from the atmosphere depends on both having enough flow and maintaining a laminar (turbulence free) flow. Turbulence from too high a gas flow rate will bring air into the area we want to shield, ruining the weld. It also wastes shielding gas.

Carbon dioxide is commonly used on mild steel for minimum cost. 75% argon + 25% carbon dioxide is used for general welding of carbon steel and low alloy steel with excellent results.

In selecting a shielding gas, consider cost, weld appearance, and weld penetration. In some cases, mixes of different gases are best for the job.

Figure 5-13 A mixture of argon and carbon dioxide shielding gas
is usually used on carbon steel and low-alloy steel

Photo courtesy of Miller Electric.

How is the welding wire transferred to the welding pool?

There are four transfer processes in MIG welding—short-circuit transfer, globular transfer, spray transfer, and pulsed-spray transfer. They are determined by the magnitude and type of welding current, the diameter of the electrode, the composition of the electrode, the electrode extension, and the shielding gas used. Going from low to high welding currents, here's how each process works.

Short-circuit transfer—This transfer occurs at the lowest current ranges and electrode diameters. It produces a small, fast-freezing weld pool suitable for joining thin sections, making open root passes, and performing out-of-position welds. Metal transfers only when it is in contact with the work at between 20 to over 200 times/second.

Globular transfer—This transfer mode occurs as the short-circuit mode transitions to the spray transfer mode. Its poor penetration and tendency to produce spatter limits its production applications.

100

Spray transfer—There are three requirements for spray transfer: argon-rich shielding gas, DCEP polarity, and a welding current above the critical value, called the transition current. Below the transition current the transfer is globular, above it transfer is by spray mechanism. Spray transfer produces a discrete stream of metal droplets that are accelerated by electrical forces to overcome gravity effects. The result is no short-circuiting, so no spatter, excellent penetration, and the ability to weld out of position. High deposition rates are possible.

Pulsed-spray transfer—Spray transfer's high deposition rate is hard to utilize on thin sections as the high currents needed to produce spray transfer often lead to burn through. The solution is to use a pulsed wave shape from the welding power supply. Not all equipment can supply this type of power, and it is mainly an industrial process. The transfer during the pulses provides the spray transfer characteristics of high deposition without spatter while the time between the pulses provides for a cooling period to prevent burn through. See Figure 5-14 and Table 5-3.

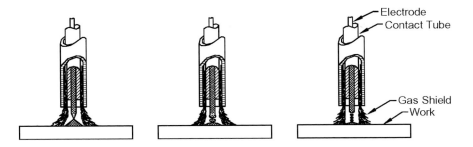

Figure 5-14 The three common types of GMAW metal transfer:
short-circuit (left), globular (middle) and spray (right)

Table 5-3 Conditions for MIG Metal Transfer and Applications

	Short Circuit	**Globular**	**Spray**	**Pulsed Spray**
Conditions for this mode	• Low current levels • Small diameter electrodes	• Current levels just above short-circuit transfer • Occurs between short circuit and spray transfer	• Occurs at above transition current • DCEP • Argon-rich shielding gas	• Same as spray but using pulsed power source
Characteristics	• Low deposition rates	• Spatter • Poor penetration	• High deposition rates • Spatter-files • Excellent penetration	• Same as spray, but can perform all positions.
Applications	• < 12 gauge • Root pass • Overhead • Flat horizontal • Vertical up • Vertical down	• > 12 gauge • Vertical down	• > 1/8 inch • Flat horizontal	• > 1/8 inch • All positions

How is MIG equipment set up?

Most constant-voltage wire feed machines are set up so that the 210 inches/ minute (5.3 meters/minute) setting will be at the ten o'clock position on the speed control or the number 3 position. Set the voltage to 18 volts. Try to weld on a scrap of the same type and thickness as the work. If the wire stubs, turn the voltage up until it does not; if the wire globs, turn the voltage down. This approach will work on a wide range of material thicknesses—22 gauge to 1/4 inch (0.7 to 6 mm) thickness. Remember, we want to set the voltage and feed rate so that welding sounds like eggs frying. Once you recognize the crackling sound of a good GMAW weld, you will find setting parameters easy.

Setup Steps

- Make sure the welding power-supply and wire feeder power are turned OFF.
- Mount the electrode wire spool on the wire feeder.
- Without letting go of the wire (it will uncoil and make a snarl if you do), open the feed roller gate, and insert the electrode wire into the feed roller mechanism and into the start of the liner tube. Close the gate, securing the wire. Make the wire feeder rolls match the wire diameter.
- Turn on the power for the wire feed motor, then use the JOG or INCH button to load the wire into the welding cable liner and up to and through the welding gun.
- Use the JOG or INCH button to actuate the electrode wire drive motor, then adjust feed roller pressure until you can no longer stop the wire feeding into the rollers by pinching the electrode wire between your thumb and index finger. Erratic wire feeding usually results from too much roller pressure, which causes the electrode wire to flatten (and loose feed) as it passes through the rollers.
- Properly secure the compressed shielding gas cylinder; crack the valve to remove any dirt. Then put the flow regulator or flow meter (or both) on the cylinder valve, and secure the other end of the gas line into the welding machine.
- Turn welding power supply ON. Using the PURGE button (or if no PURGE button, squeezing the welding gun button being sure the gun is not touching anything or anybody) for several seconds to remove air from the lines and fill them with shielding gas. While performing this purge, set the flow meter or flow regulator for the recommended purge gas flow rate; 20 ft^3/hour (9.5 l/min) for 0.035 inch (0.9 mm) wire would be a good starting point. On larger diameter electrode wire, begin with 20 ft^3/hour (9.5 l/min) on horizontal and 30 to 35 ft^3/hour (14 to 16.5 l/min) on out-of-position welds.
- Set proper polarity (DCRP), voltage, and wire feed speed. Get this information from manufacturer's data sheets for the electrode wire or from tables attached to the welding machine. See Figure 5-15.

- Attach the work lead to a clean spot on the work.
- Use wire cutters to trim the electrode stickout to proper length—about 1/2 inch (12 mm). See Figure 5-16.
- Make sure welding area is dry and free of flammable materials as well as volatile fumes and that others present are protected from arc radiation and sparks.
- Put on your welding helmet (your safety glasses should already be on and remain on under the helmet), position the electrode against the work, squeeze the trigger, and begin welding.

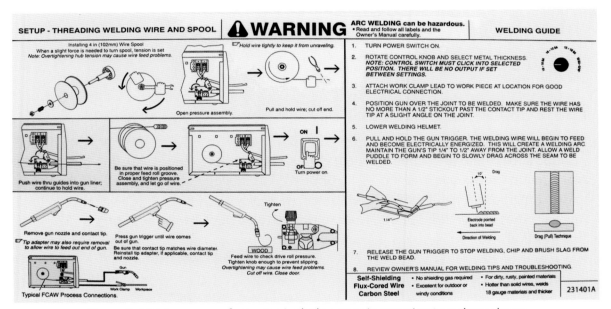

Figure 5-15 Many manufacturers include setup instructions on door charts

Photo courtesy of Hobart Welders.

Figure 5-16 Use MIG pliers or a pair of side cutters to trim the electrode to the proper length before welding

Setting the Arc Voltage

We really want to control arc length, but it is hard to measure and closely related to arc voltage, so we choose to control arc voltage. Most welding specifications indicate arc voltage setting, not arc length. This is because with all other variables held constant, arc length is proportional to arc voltage. Arc voltage depends on materials welded, shielding gas, and metal transfer mode.

Too short an arc may produce stubbing—cold wire pushes into the work-piece upsetting the smooth flow of shielding gas and pumping air into the arc stream, causing porosity and cracking from atmospheric nitrogen.

Too long an arc can cause the arc to wander, degrading both penetration and weld bead quality. In general, higher voltages tend to flatten the weld bead and increase the width of the fusion zone. Very high arc voltages can cause porosity, undercut, and spatter.

The best way to set arc voltage is to begin at the recommended voltage based on welding tables from AWS or the electrode manufacturer's data sheets. The base metal, type of metal transfer, and shielding gas will yield a starting point. Exact refinement will come from several trial runs.

For short-circuit transfer you should set the common industrial (CV) constant voltage (constant potential) power supply at 18 volts and your wire feed speed at 210 inches per minute as a beginning setting: then observe what is happening to the weld area. If:

• welding wire coming from the contact tip melts but the metal you are welding is not melting, increase the voltage.

• welding wire is coming out too fast and pushing into the material (stubbing), increase your voltage setting a notch at a time, do not increase a full number until the short-circuiting begins and the sound of the welding is similar to the sound of bacon frying.

• welding wire forms droplets and splashes into the molten pool, decrease your voltage.

Determining Wire Feed Rate

On many wire feed machines, wire feed rate is marked on the speed control or by a digital read-out, but the best way to check the speed is to run out electrode (being careful not to permit an arc) for one minute, then measure the wire output. Usually, each number on a wire feed speed adjustment scale represents an increment of 70 inches/minute (1.8 m/minute). In general, both the wire feed rate and the arc voltage will need adjustment to get the optimum weld.

What are some joint preparation steps for MIG welding?

Metals from 0.005 to 3/16 inch (0.130 mm to 4.8 mm) can be welded without edge preparation.

Metals from 0.062 inches to 3/8 inch (1.6 to 10 mm) can be welded in a single pass with joint edge preparation.

Multipass welding is required above 3/8 inch (10mm) and preparation is needed; there is no limit to the thickness of metal that can be welded.

Because the MIG electrode has a smaller diameter than the SMAW (stick welding) electrode, it can get to the bottom of a narrower V-groove. Therefore, the preparation for V-grooves may be made to a smaller angle. This saves as much as 50% on welding filler metal and welding time.

There is no problem using an SMAW V-groove design for GMAW; it will just take longer than necessary. See Figure 5-17.

105

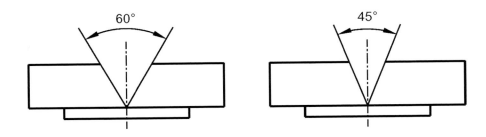

Figure 5-17 V-groove joint for stick welding (left) and MIG welding (right)

How does the orientation of the electrode during welding affect the weld?

How the gun is held affects the weld bead:

In the *forehand technique* with a lead angle, the gun is tipped toward the direction of welding. Going from perpendicular to a lead angle gives less penetration and a wider, thinner weld. On some materials, particularly aluminum, a lead technique will produce a cleaning action just ahead of the weld pool. This promotes wetting—the visible inter-melting and fusion of the wire feed electrode with the base metal—and reduces base metal oxidation.

In the *torch technique*, the gun is perpendicular to the direction of welding. It has neither a lead nor a drag angle.

In the *backhand technique* with a drag angle, the gun is tipped in the direction of welding. Maximum penetration occurs in the flat welding position with a drag angle of 25° from perpendicular. For all other positions, the backhand technique is used with a drag angle of 5 to 15°.

See Figure 5-18 for comparison of these three positions and resulting bead width and penetration.

106

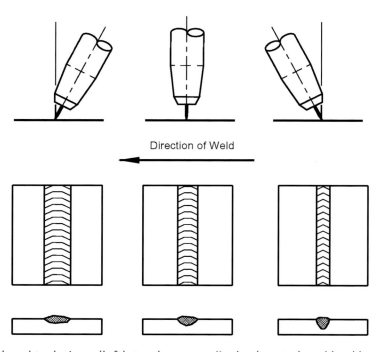

Direction of Weld

Figure 5-18 Forehand technique (left), torch perpendicular (center) and backhand technique (right) and their results

The Effects of Weld Position

Spray-metal transfer is usually done in the flat position because it is easier to control a large weld pool when flat.

Pulsed and short-circuit metal transfer may be done in any position.

Welding fillets done with spray transfer in the flat position are more uniform than the same welds done in the horizontal position.

Welding electrode wire 0.045 inches (1.2 mm) or less is usually used in the overhead and vertical positions to make weld pool control easier by making the pool smaller.

Sometimes welding downhill at about 15° can increase welding speed, decrease penetration, and reduce weld thickness. This tactic is helpful in welding thinner sheet metals.

Table 5-4 Effects of changing weld parameters

Welding Variable ↓	Result Wanted							
	Penetration		Deposition Rate		Bead Size		Bead Width	
	✳	❋	✳	❋	✳	❋	✳	❋
Wire Feed Rate (Current)	✳	❋	✳	❋	✳	❋	⇔	⇔
Voltage*	Little Effect	Little Effect	⇔	⇔	⇔	⇔	✳	❋
Travel Speed	Little Effect	Little Effect	⇔	⇔	❋	✳	✳	❋
Stickout	❋	✳	✳	❋	✳	❋	❋	✳
Wire Diameter	❋	✳	❋	✳	⇔	⇔	⇔	⇔
Electrode Orientation	Back Hand	Fore Hand	⇔	⇔	⇔	⇔	Back Hand	Fore Hand
Shield Gas %CO_2	✳	❋	⇔	⇔	⇔	⇔	✳	❋

✳ = Increasing ❋ = Decreasing ⇔ = No Effect

* Voltage settings can be important for sheet metal.

What are typical MIG welding problems and solutions?

Table 5-5 provides some good starting points for troubleshooting, but it is by no means comprehensive.

Table 5-5 MIG Troubleshooting		
Problem	**Possible Cause**	**Remedy**
Undercutting	1. Travel speed too high 2. Welding voltage too high 3. Excessive current 4. Insufficient dwell 5. Gun angle	1. Use slower travel speed 2. Reduce voltage 3. Reduce wire feed speed 4. Increase dwell at puddle edge 5. Change gun angle so arc force helps in placing metal

Porosity	1. Inadequate shielding gas coverage
	
	2. Gas contamination
	3. Electrode contamination
	
	4. Workpiece contamination
	
	5. Arc voltage too high
	6. Excess contact tube-to-work distance
		1. Gas flow too high or too low
		  Sheild work from drafts
		  Decrease torch to work distance
		  Hold gun at end of weld until metal solififies
		2. Use welding grade shielding gas
		3. Use clean and dry electrodes
		4. Remove all dirt, rust, paint, moisture from works surface
		5. Reduce voltage
		6. Reduce stickout

Incomplete Fusion	1. Weld zone surfaces not clean 2. Insufficient heat input 3. Too large a weld a puddle 4. Improper weld technique 5. Improper joint design 6. Excessive travel speed	1. Clean weld surfaces carefully 2. Increase wire feed speed and voltage 3. Reduce weaving, increase travel speed 4. When weaving, dwell momentarily at edge of groove 5. Use joint design with wide enough angle to all access to groove bottom 6. Reduce travel speed

Table 5-5 (cont.) MIG Troubleshooting		
Problem	**Possible Cause**	**Remedy**
Incomplete Joint Penetration	1. Weld zone surfaces not clean 2. Insufficient heat input 3. Too large a weld a puddle 4. Improper weld technique 5. Improper joint design 6. Excessive travel speed	1. Clean weld surfaces carefully 2. Increase wire feed speed and voltage 3. Reduce weaving, increase travel speed 4. When weaving, dwell momentarily at edge of groove 5. Use joint design with wide enough angle to all access to groove bottom 6. Reduce travel speed
Excess Melt-through	1. Excessive heat input 2. Improper joint penetration	1. Reduce wire feed speed 2. Reduce root opening Reduce root face dimension

Weld Metal Cracks	1. Improper joint design
	
	2. Too-high a weld depth-to-width ratio
	3. Too-small a weld bead (particularly in fillet and root beads)
	4. Hot shortness
	
	5. High restraint of joint members
		1. Joint design to provide adequate weld metal to overcome constraint conditions
		2. Decrease arc voltage, decrease wire feed speed or both
		3. Decrease travel speed
		
		4. Use electrodes with higher manganese content
		   Increase groove angle to increase filler metal in weld
		   Change filler metal
		5. Use preheat
		   Adjust welding sequence

Heat Affected Zone (HAZ) Cracks	1. Hardening in the HAZ
	2. Residual stresses too high
	3. Hydrogen embrittlement
		1. Preheat to retard cooling rate
		2. Use stress relief heat treatment
		3. Use clean electrode
		   Use dry shielding gas
		   Hold weld at elevated temperature for several hours to permit diffusion of hydrogen out of weld

109

MIG Welding Safety

- Protect face and eyes from sparks and radiation with a helmet and lens of appropriate density number. Note that the heat of MIG welding can be great enough to crack a filter lens, and for this reason, all welding masks should have an appropriate glass or plastic filter cover both in front of and behind the welding glass filter.

- The total radiated energy both visible and invisible of MIG is much higher than SMAW processes, so extra precautions must be taken to protect the eyes and skin. Also, because the GMAW process produces less smoke than the SMAW process, more of the radiation produced is available to harm the welder. To protect yourself, use the Table 5-6 to determine the filter glass shade to use on GMAW: try the darker shade for the current you are using, and drop to the next lighter shade until you can see the welding action clearly. Never drop to a shade lighter than the lowest recommended one. Protect your skin from ultraviolet, invisible radiation with dark leather or wool clothing. Pay particular attention to both the direct and reflected light from the arc on the arms (cover with a shirt or leathers), the neck area (add a commercially available leather flap to the bottom of your welding helmet) and the top of your head (wear a welder's cap). See Table 5-6.

Table 5-6 Lens shade selection chart for GMAW and FCAW

Welding Current (A)	Minimum Protective Shade	Suggested Shade for Comfort
> 60	7	—
60-160	10	11
160-250	10	12
250-500	10	14

- GMAW can be a noisy process, and hearing protection may be necessary for both comfort and safety. Continuous exposure to relatively low noise levels (especially high frequencies) can cause permanent hearing loss. Ear plugs and ear muffs may be needed.

- Wear safety glasses to protect against silicon popping off the weld surface and other hazards.

How does the flux cored arc welding (FCAW) process work?

The structure and chemical composition of FCAW electrode wire makes the difference between GMAW and FCAW. Unlike GMAW, the FCAW electrodes consist of a thin-walled metal tube filled with flux, not a solid wire. The powdered flux provides alloying elements, arc stabilizers, denitriders, deoxidizers, slag formers, and shielding gas generating chemicals. For many electrodes, the volume and forcefulness of the shielding gas eliminates the need for external shielding gas. This is called self-shielded and has the AWS designation FCAW-S. If a shielding gas is used, it is called gas-shielded and carries the designation FCAW-G. The slag produced also shields the weld pool from atmospheric oxygen and nitrogen and retards its cooling rate to reduce martensite formation. Working in tandem, the shielding gas and slag allow FCAW to be used successfully in field conditions. Because of the FCAW-S heavy smoke, no physical shielding is needed to protect the welding from wind. Depending on the electrode, both DCEP and DCEN are used; AC is not. See Figures 5-19 and 5-20.

Only ferrous metals and nickel-based alloys: all low- and medium-carbon steels, some low-alloy steels, and stainless steels, can be welded with FCAW.

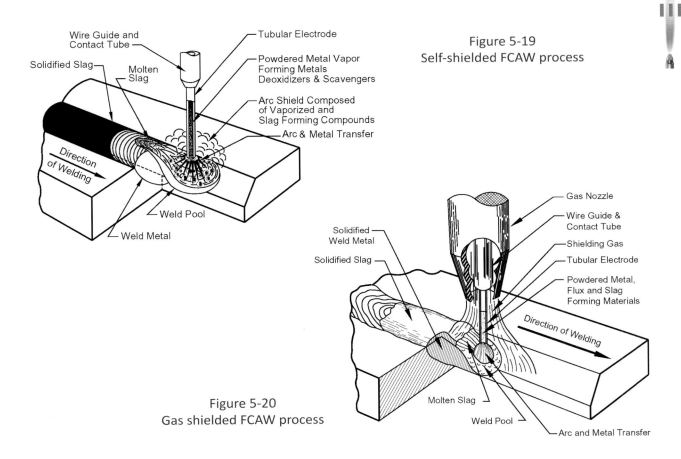

Figure 5-19
Self-shielded FCAW process

Figure 5-20
Gas shielded FCAW process

Pros and Cons of FCAW

Advantages

- Because of its excellent penetration, FCAW can use groove angles as narrow as 30°, saving as much as 50% of the filler metal needed for SMAW, which requires larger groove angles. In some cases, FCAW can avoid joint beveling in metal up to 1/2 inch (13 mm) thick.

- Different thickness material can be welded with the same electrode by power supply adjustment.

- Easier to use than SMAW.

- Excellent weld pool control.

- FCAW has better welder visibility of the weld pool than GMAW because the gas diffuser nozzle is not needed and can be removed.

- FCAW is often permitted by codes for critical welds on boilers, pressure vessels, and structural steel.

- FCAW oxidizers and fluxing agents permit excellent quality welds to be made on metals with some surface oxides and mill scale. Often metal from flame cutting can be welded without further preparation, a major cost saving.

- High deposition rates of more than 25 lb/hour (11.3 kg/hour) are possible compared with 10 lb/hour (4.5 kg/hour) for SMAW with 1/4- inch (6 mm) diameter electrodes.

- Self-shielded FCAW electrodes work better in windy field conditions than gas-shielded GMAW electrodes.

- There is no stub loss as in SMAW where losses can average 11% of the electrode.

- Unlimited thicknesses can be joined with multiple passes.

- Welds can be made in all positions.

- Wire feeder and welder must be close to the point of welding.

Disadvantages

- Slag removal requires an additional step, similar to SMAW.

- FCAW generates large volumes of fumes and smoke, requiring additional ventilation indoors and reducing welder visibility during the process.

112

What equipment is needed for an FCAW welding outfit?

- FCAW requires a constant-voltage (CV) power supply, the same as GMAW.

- Depending on the welding electrode wire, external gas shielding from a cylinder/flow regulator/flow meter may or may not be needed. Using an external shielding gas when the electrode is designed to work without one can produce a defective weld, so be sure to consult the manufacturer's data sheets.

- Wire feeders for larger diameter electrodes usually have four feed rollers instead of two. This reduces the feed roller pressure on the electrode to avoid damage; FCAW electrode wire is more fragile than the solid GMAW electrode wire.

- Because very high welding currents are used—they can exceed 250 amperes—many FCAW guns have a sheet metal heat shield to protect the operator's hands from the intense heat of the weld. Although most are air-cooled, some FCAW guns are water-cooled to operate on a 100% duty cycle. See Figure 5-21.

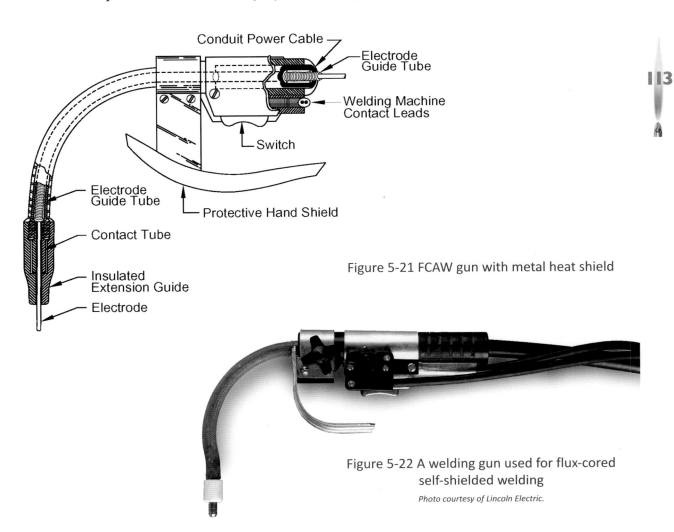

Figure 5-21 FCAW gun with metal heat shield

Figure 5-22 A welding gun used for flux-cored self-shielded welding

Photo courtesy of Lincoln Electric.

How do you select FCAW electrodes?

See Figure 5-23.

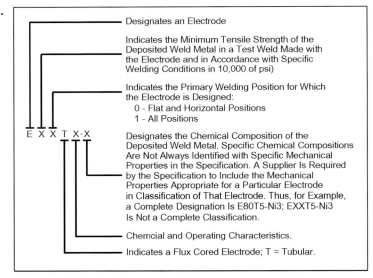

Figure 5-23 FCAW wire classification for mild and low alloy steel

Table 5-7 shows which electrodes require gas shielding.

Table 5-7 Shielding and Polarity Requirements for Mild Steel FCAW Electrodes		
AWS Classification	**External Shielding Meduim**	**Current and Polarity**
EXXT-1 (Multiple-Pass)	CO_2	DCEP
EXXT-2 (Single-Pass)	CO_2	DCEP
EXXT-3 (Single-Pass)	None	DCEP
EXXT-4 (Multiple-Pass)	None	DCEP
EXXT-5 (Multiple-Pass)	CO_2	DCEP
EXXT-6 (Multiple-Pass)	None	DCEP
EXXT-7 (Multiple-Pass)	None	DCEN
EXXT-8 (Multiple-Pass)	None	DCEN
EXXT-9 (Multiple-Pass)	None	DCEN
EXXT-10 (Single-Pass)	None	DCEN
EXXT-11 (Multiple-Pass)	None	DCEN
EXXT-G (Multiple-Pass)	*	*
EXXT-GS (Single-Pass)	*	*

*As agreed upon betrween supplier and user.

What are typical FCAW problems and solutions?

Table 5-8 FCAW Troubleshooting

Problem	Possible Cause	Corrective Action
Porosity	1. Low gas flow	1. Increase gas flow setting Clean spatter-clogged nozzle
	2. High gas flow	2. Decrease to eliminate turbulence
	3. Excessive wind drafts	3. Shield weld zone from wind or draft
	4. Contaminated gas	4. Check gas sources Check for leak in hoses and fittings
	5. Contaminated base filler	5. Clean weld joint faces
	6. Contaminated filler wire	6. Remove drawing compound on wire Clean oil from rollers Avoid filler wire Rebrake filler wire
	7. Insufficient flux in core	7. Change electrode
	8. Excessive voltage	8. Reset voltage
	9. Excess electrode stickout	9. Reset stickout & current
	10. Insufficient electrode stickout	10. Reset stickout & current
	11. Excessive travel speed	11. Adjust speed
	12. Excessive voltage	12. Reset voltage
	13. Insufficient electrode stickout (self-shielded electrodes only)	13. Reset stickout and balance current
	14. Excessive travel speed	14. Reset stickout and balance current
Incomplete fusion or Penetration	1. Improper manipulation	1. Direct electrode to joint root
	2. Improper parameters	2. Increase current Reduce travel speed Decrease stickout Reduce wire size Increase travel speed (self-shielded electrodes)
	3. Improper joint design	3. Increase root opening Reduce root face
Cracking	1. Improper manipulation	1. Reduce restraint Preheat Use more ductile metal Employ peening
	2. Improper electrode Insufficient deoxidizers or inconsistent flux fill in core	2. Check formulation of flux
Electrode feeding	1. Excessive conact tip wear	1. Reduce drive roll pressure
	2. Melted or stuck contact tip	2. Reduce voltage Replace worn liner
	3. Dirty wire conduit in cable	3. Change conduit liner Clean out with compressed air

115

How is FCAW equipment set up?

• Same as GMAW.

• Be sure to consult the electrode manufacturer's recommendations for polarity (most are DCEN), wire feed speed, and voltage setting.

• If a shielding gas is needed, it is likely to be carbon dioxide; the most common mixture of gases is 75% argon and 25% carbon dioxide.

FCAW Safety

• Same as MIG except needs better ventilation indoors since FCAW generates more smoke and fumes—about the same amount as SMAW.

• Lens shade same as MIG. See Table 5-6.

116

MIG Welding Tips from West Coast Customs

• Keep equipment in good working order.

• Keep bends out of the MIG hose.

• Prepare all metal properly. Finish prep work by wiping down the metal using acetone. The cleaner the metal, the cleaner the weld.

• Never let the equipment sit outside because the electrode wire can become contaminated.

• Stay relaxed when welding. The more relaxed you are while welding, the better the results. Hold the gun with one hand, and use the other hand to support the gun hand.

Because the electrode, filler rod, and shielding gas are all contained within the nozzle of the welding gun, making the process much easier to master than other types of welding. It also means you can use two hands to hold the gun steady. You can use either the forehand method, where the gun is pushed in the direction of travel, or the backhand method, where the gun is dragged in the direction of travel.

117

Figure 5-25 Pulling the gun along the joint gives the welder a clear view of the joint. Hold the gun at about a 35 degree angle for best results

Figure 5-24 After adjusting the stickout, squeeze the trigger and touch the electrode to the work to start. The welder used tack welds to keep this tee joint stable. Try to maintain a contact tip to work distance of 3/8 to 1/2 inch. Make smaller ovals with the tip as you move forward

Figure 5-26 Notice the consistent ridges and how the bead is even on both sides of the seam

Gas Tungsten Arc Welding

TIG Welding

G as Tungsten Arc Welding (GTAW) is commonly called *TIG* welding from tungsten inert gas welding. In this process, the arc produced through a tungsten electrode melts the base metal. The heat is so intense and the arc so focused that extremely accurate and fine welds can be produced. Unlike MIG, or Gas Metal Arc Welding, discussed in chapter 5, the tungsten electrode is not consumed nor does it become part of the weld. This type of welding requires a shielding gas.

How does the process work?

What is the basic equipment setup

for TIG Welding?

How are the electrodes selected?

What shielding gases are used with TIG Welding?

What is the TIG Welding setup process?

What are some problems and solutions

for TIG Welding?

TIG welding tips from West Coast Customs

How does the process work?

A low-voltage, high-current, continuous arc is formed between a tungsten electrode on the welding torch and the work through an inert gas, either argon or helium. The intense heat of this arc, approximately 10,000°F (5,500°C), melts the surface of the base metal forming the weld pool. On thinner metals, edge joints, and flange joints, no metal is added; this is called *autogenous welding*. On thicker materials, filler metal is added in the form of a wire or rod fed into the arc. The inert gas supplied through the welding torch not only provides a suitable arc, it displaces air, shielding the weld pool and the electrode from atmospheric contaminants. No metal is transferred across the arc, so there is no spatter and little or no smoke. Since the welder may continuously control welding current with the fingertip or foot control, there is excellent control of the weld pool resulting in high-quality welds.

Pros and Cons of TIG welding

Advantages
- Welds are of high quality.
- Welds nearly all metals and alloys.
- All weld positions are possible.
- No slag developed.
- Excellent welder visibility of arc and weld pool.
- Little post-weld cleaning needed.
- There is no spatter.
- Provides excellent welder control of root pass weld penetration.
- Allows heat source and filler metal to be controlled independently.
- Joins dissimilar metals.

Disadvantages
- Higher welder skills are required than other welding processes.
- Lower deposition rate and productivity compared with other processes.
- Equipment is more complex and expensive than other more productive processes.
- Low tolerance for contaminants in filler or base metals.
- Possible problems welding in drafty environments.

What is the basic equipment setup for TIG welding?

- Constant-current (CC) welding power supply, either DC or AC/DC
- Tungsten electrodes
- GTAW torch and associated cable
- Shielding gas, cylinder, regulator, or flow meter and hoses
- Work clamp and lead
- Filler metal (optional)
- Foot pedal or finger-tip power control (optional)
- Cooling water source and drain (optional)
- Cooling water recirculating system (optional)

See Figure 6–1.

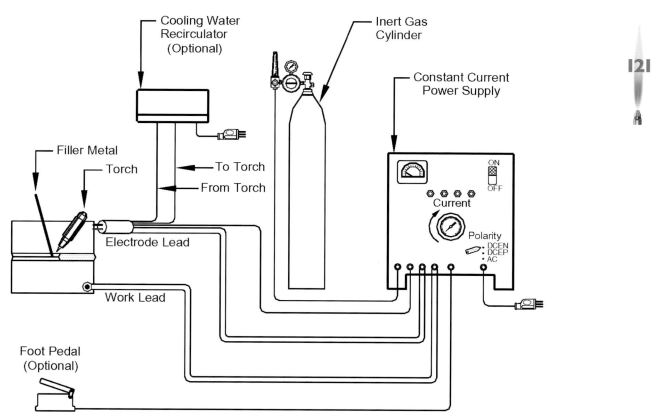

Figure 6–1 TIG Welding equipment

Figure 6-2 Manufacturers supply complete
TIG kits.
Note the foot pedal in the background

Photo courtesty of Miller Electric.

Torches and Cables

The three principal torch designs are

- Small, gas-cooled manual welding torches rated up to 200 amperes, see Figure 6-3. The relatively cool inert shielding gas on its way to the arc cools the torch.

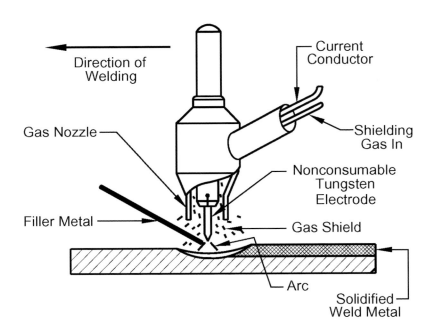

Figure 6-3 Gas-cooled TIG torch

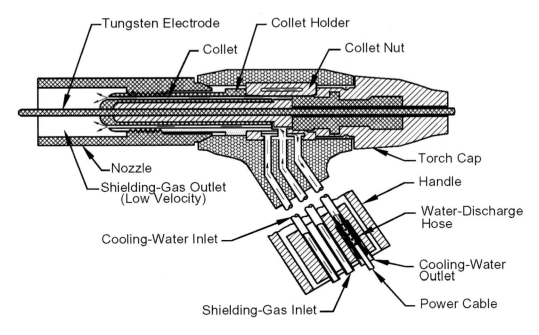

Figure 6-4 Water-cooled TIG torch

- Large water-cooled manual welding torches rated over 200 amperes, see Figure 6-4. Either water from a tap or a recirculating cooler prevents heat from destroying the torch.

- Automatic welding torches are similar in the cooling mechanics but the configuration is different. See Figure 6-5.

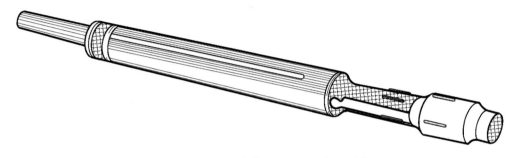

Figure 6-5 TIG torch for automatic welding

Figure 6-6 Here is a typical manual TIG torch. Note that the electrode is mounted perpendicular to the incoming cables.

Photo courtesy of Lincoln Electric.

Generally, manual torches have the electrode mounted at an angle to the cables to relieve the welder of having to support the twisting force of cables coming out of the back of the torch. See Figure 6-6. However, some jobs are more easily done with pencil-style or straight-line torches where the cables come directly out of the back of the torch like those designed for automatic applications. These smaller, pencil-style manual designs usually do not have water cooling.

Automatic torches also have a straight-through design and are water cooled for 100% duty cycle use. Cooling water may come from a tap, be used once, and discharged to a drain. Another alternative is to use an evaporative cooler with a circulating pump to provide cooling water. This permits welding away from water lines and drains and is also more economical. The heaviest torches are rated at 600 amperes.

All torches have precision copper collets to grip and center the electrode in the torch. The collets transfer welding current to the tungsten electrode from the welding cable. They also remove heat from the electrode to keep it from melting. There are holes or ports surrounding the electrode holder to distribute inert shielding gas evenly around the electrode and over the weld pool.

Insulating Nozzles

The nozzles are made from ceramic, metal, alumina, or fused quartz, which provides see-through visibility. They can withstand both shock and intense heat. Their function is to flow gas around the electrode and into a stream at the weld pool. Their inside diameter is measured in sixteenths of an inch. If a nozzle is described as a number 5, its inside diameter is 5/16 inch, a number 10 is 10/16 or 5/8 inch inside diameter. The inside diameter of the nozzle should be at least three times the electrode diameter.

125

Figure 6-7 Here is an example of a TIG torch kit. The pink cups are nozzles, the brass items are collets and collet bodies.

Photo courtesy of Lincoln Electric.

Figure 6-8 GMAW equipment can handle all types of metal welding jobs.

Photo courtesy of Hobart Welders.

Electricity and the Arc

The voltage used in GTAW ranges from 10 to over 40 volts; current ranges from 1 to over 1000 amperes. The TIG arc is shaped much like an inverted funnel: nearly a point at the electrode and flaring out to a circle on the flat side of the work metal. Arc length is roughly proportional to arc voltage. Generally arc length runs from one to four times electrode diameter.

Why Doesn't the Tungsten Electrode Melt?

While a very small amount of the electrode does melt and ends up in the weld, the electrode resists melting because:

- Tungsten has the highest melting point of all metals 6,170°F (3,420°C).
- Tungsten has high thermal conductivity so heat can readily flow from the electrode's hot tip to the cooler collet.
- Torch collets are designed to remove heat from the electrode, and the torches themselves are gas or water-cooled.

Polarities Used in TIG Welding

DCEN (direct current electrode negative)—consider a torch connected to DCEN. When the arc begins, the electrode metal heats and as a result has enough energy to release electrons from its surface. This process is called *thermionic emission.* The voltage between the electrode and the work exerts a strong pull drawing the cloud of electrons around the surface of the electrode toward the positive work surface and accelerating the electrons to high speed. As the electrons move through the electrically excited ionized gas, they generate heat from their friction with the inert gas atoms. When these high-speed electrons strike the work surface, their kinetic energy becomes heat. Over 99% of the current is by electron flow, the remainder by positive ions. That is, base metal ions stripped of one or more electrons become positively charged as a result and produce a current flow as they move from the work to the electrode, the direction opposite to electron flow. DCEN produces the most heat and deep weld penetration within a narrow area. However, it does not provide cleaning action, called *cathodic etching* that is necessary for welding aluminum and magnesium.

DCEP (direct current electrode positive)—in this polarity, electrons leave the work and accelerate as they make their way through the ionized gas to the electrode under the influence of the voltage across the arc. Their impact on the electrode generates intense heat that will melt the electrode if not removed. Because this electron bombardment produces more heat at the electrode with DCEP than DCEN, a larger electrode must be used running the same current level to absorb the additional heat. In fact, DCEP can handle only 10% of the current the same sized electrode can run with DCEN. While DCEP produces less penetration than DCEN, it provides a wider area of heating than DCEN. It also provides a cleaning effect on the base metal

127

Current Type	DCEN	DCEP	AC (Balanced)
Electrode Polarity	Negative	Positive	
Electron and Ion Flow / Penetration Characteristics			
Oxide Cleaning Action	No	Yes	Yes-Once Every Half Cycle
Heat Balance in the Arc (Approx.)	70% at Work End / 30% at Eectrode end	30% at Work End / 70% at Electrode End	50% at Work End / 50% at Electrode End
Penetration	Deep, Narrow	Shallow, Wide	Medium
Electrode Capacity	Excellent (For Example 1/8" (3.2 mm) 400A	Poor (For Example 1/4" (6.4 mm) 120A	Good (For Example 1/8" (3.2 mm) 225A

Figure 6-9 Characteristics of current types for GTAW

within the arc area. Positive ion bombardment onto the work produces cathodic etching. This cleaning effect is critical when welding metals like aluminum and magnesium that instantly form surface oxides when in contact with the atmosphere. The cleaning effect removes these oxides, and GTAW's inert gas blanket prevents them from reforming.

AC—provides equal heat at the electrode and the work because each receives electron bombardment half the time. It provides penetration midway between DCEN and DCEP. It is used most often to weld aluminum and magnesium because its cleaning effect is essential. See Figure 6-9 on previous page.

Unbalanced AC—provides enough cleaning effect to perform welding on aluminum and magnesium, but uses more of the cycle to put heat in the work. Unbalanced AC attempts to get the best of DCEN and DCEP. See Figure 6-10.

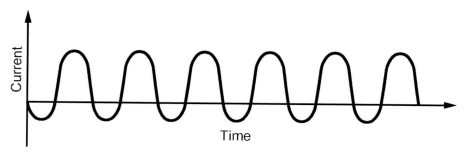

Figure 6-10 Unbalanced AC GTAW waveform to provide surface cleaning effect.

Unbalanced pulses—Similar to unbalanced AC in providing a cleaning effect on aluminum and magnesium, but has faster rise and fall times for a smoother arc. See Figure 6-11.

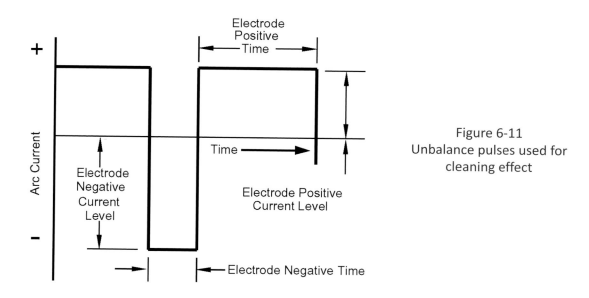

Figure 6-11
Unbalance pulses used for cleaning effect

Pulse waveforms with ramps up and ramps down—these provide a gradual starting current (the *upslope*), a stable welding current, and a gradual tapering off of the current when stopping (the *downslope*). The downslope allows the final weld pool to be completely filled and not leave a crater. Note the term *slope* as used here has nothing to do with the slope of a constant-current power supply; it refers to the shape of the current waveform when starting and stopping.

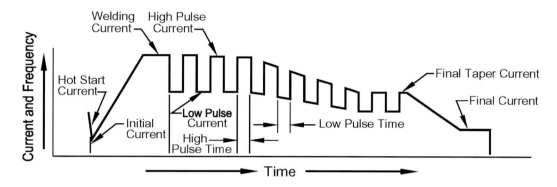

Figure 6-12 GTAW pulsed current waveform with ramps up and down

129

Other Energy Sources

High-frequency (several Megahertz), high-voltage, low-current energy is used in addition to the low-frequency, low-voltage energy supplied to the arc. On DC polarity, high-frequency is used only to start the arc; the power supply automatically shuts off the high-frequency after the arc is established. With AC polarity, high-frequency energy is supplied continuously to maintain the arc as the polarity reverses or the arc would extinguish.

Because many GTAW power supplies use high-frequency, high-voltage pulses to start or maintain the welding arc, these pulses can cause interference with radio reception over a wide area if the systems are not properly installed, grounded, and maintained. Not only can these pulses disrupt radio and TV reception, they can interfere with critical police, fire, and aircraft communications. For this reason, the FCC has established rules limiting the level of radiated power from welding machines.

How are the electrodes selected?

For a look at the seven classes of tungsten electrodes in AWS specifications, see Table 6–1.

Table 6–1 AWS tungsten TIG electrode classification

AWS Class.	Composition	Color Code
EWP	Pure tungsten	Green
EWCe-2	97.3% tungsten, 2% cerium oxide	Orange
EWLa-1	98.3% tungsten, 1% lanthanum oxide	Black
EWTh-1	98.3% tungsten, 1% thorium oxide	Yellow
EWTh-2	97.3% tungsten, 2% thorium oxide	Red
EWZr-1	99.1% tungsten, 0.25% zirconium oxide	Brown
EWG	94.5% tungsten, remainder not specified	Gray

PureTungsten Electrode (EWP). Pure tungsten electrodes contain a minimum of 99.5 wt-% tungsten, with no intentional alloying elements. The current-carrying capacity is lower than the alloyed tungsten electrodes. Pure tungsten electrodes are used primarily with alternating current (AC) for welding aluminum and magnesium alloys. Using AC, the tip of the EWP forms a clean balled end, which provides good arc stability. EWP may also be used with direct current (DC), but they do not provide the arc initiation and arc stability of other alloyed electrodes.

2% Cerium Oxide (EWCe-2) Electrodes. The cerium oxide tungsten electrodes (EWCe-2) are tungsten alloyed electrodes which contain a nominal 2 wt-% cerium oxide. These electrodes were developed as a possible replacement for the thoriated tungsten electrodes because ceria, unlike thoria, is not radioactive. Ceriated tungsten electrodes, when compared to pure tungsten provide similar current levels but with improved arc starting and arc stability characteristics like thoriated tungsten electrodes. They also tend to have last longer than thoriated tungsten electrodes. They will successfully operate either AC or DC.

Lanthanated Tungsten Electrodes (EWLa-1). Lanthanated (EWLa-1) tungsten electrodes were developed for the same reason the ceriated tungsten electrodes were developed; lanthana is not radioactive. These electrodes contain a nominal 1 wt-% lanthanum oxide. The current levels and operating characteristics are very similar to the ceriated tungsten electrodes.

Characteristics of 1% (EWTh-1) or 2% Thorium Electrodes (EWTh-2). The EWTh-1 or 1% thorium oxide tungsten electrode contain a nominal 1wt% thoria and the EWTh-2 contain a nominal 2wt-% thoria evenly dispersed throughout their length. Thoriated tungsten electrodes are used because the electrodes thoria provides higher electron emission, allowing increased current-carrying capacity about 20% higher than pure tungsten electrodes. Thoria also reduces electrode tip temperatures and provides greater resistance to contamination of the weld. These electrodes provide easier arc starting and the arc is more stable than with pure tungsten electrodes or zirconiated tungsten electrodes when using DC.

Zirconiated Tungsten Electrodes (EWZr-1). The zirconiated tungsten electrodes (EWZr-1) contain a small amount (0.15 to 0.40 wt-%) of zirconium oxide. These electrodes have welding characteristics that fall between those of pure tungsten and thoriated tungsten electrodes. They are normally the electrode of choice for AC welding of aluminum and magnesium alloys because they combine the desirable arc stability characteristics and balled end typical of pure tungsten and have a higher current carrying capacity and better arc starting characteristics of thoriated tungsten electrodes. Zirconiated tungsten electrodes are more resistant to tungsten contamination of the weld pool than pure tungsten and are preferred for radiographic quality welding applications.

131

Common Electrode Sizes

In inches they are:

0.010	0.020	0.040	0.060	1/16
3/32	1/8	5/32	3/16	1/4

Preparing Electrodes

Tungsten electrodes are color coded as shown in Table 6-1. The coloring may wear off, and once it does, all electrodes look pretty much the same, so keep them organized in labeled containers. Before using, the electrodes they must be sharpened. Use a special chemical sharpener or a grinding wheel.

Figure 6-13 Sharpen on a dedicated grinding wheel. Sharp tips make extremely fine work possible with TIG welding.

When grinding, follow the manufacturer's directions and use the following precautions:

- Grind with the axis of the electrode perpendicular to the face of grinding wheel. Do not grind on the side of the grinding wheel.
- Reserve the grinding wheel for electrodes only so that other metals will not contaminate the tip.
- Use exhaust hoods when grinding thoriated tungsten to prevent breathing in the dust.

Shapes of Electrode Tips

A more pointed tip produces a more directional and stiffer arc. Many welding procedures call for a specific shape tip. Electrons leave a tapered tip more easily than a blunt one. The basic shapes are:

- Blunt
- Tapered with balled end—A ball forms on a pure tungsten tip when it melts.
- Tapered—A ground point can be maintained on a thoriated tungsten tip for some time before erosion blunts it.
  See Figure 6-14.

132

Figure 6-14 Electrode shapes: blunt (left), tapered with balled end (center) and tapered (right).

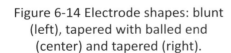

What shielding gases are used with GTAW?

Helium—a mono-atomic inert gas is extracted from natural gas by distillation. Welding helium has a minimum purity of 99.99%.

Argon—is also a mono-atomic inert gas extracted from the atmosphere by distillation of liquid air. Welding-grade argon has a minimum purity of 99.95%, which is fine for most metals except reactive and refractory ones which need 99.997% purity.

Argon versus Helium

Argon is used more frequently than helium in TIG because of these advantages:
- Smoother and more stable arc action
- Reduced penetration
- Cleaning action on aluminum and magnesium
- Lower cost and greater availability than helium (which comes only from natural gas wells in the US)
- Lower argon flow rates than helium provide adequate shielding
- Better resistance to cross-drafts because it is heavier than air
- Easier arc starting than with helium

Argon has an atomic weight of 40 while helium has an atomic weight of just 4. This makes helium much lighter than air and has the tendency to float up and away from the weld pool instead of blanketing it. Argon is closer in weight to air, so has better blanketing properties.

However, helium provides much better penetration than argon and is often used to join metals with high thermal conductivity or to make joints on thick sections. Sometimes mixtures of both helium and argon are used to get some of the best characteristics of both gases. Helium works better welding aluminum and magnesium.

133

Containing Shielding Gas Near the Joint

Some ways GTAW shielding gas can be contained near the joint being welded are illustrated in Figure 6-15.

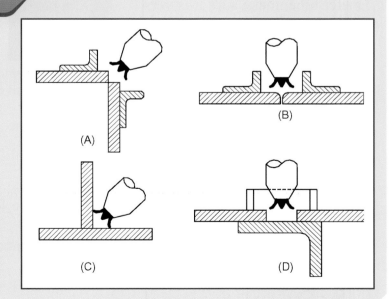

Figure 6-15 Barriers used to contain shielding gas.

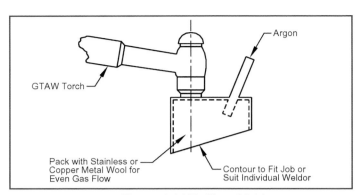

TOOL TIP

Trailing Shields

For metals like titanium, a trailing shield is necessary to keep air away from the molten metal until it has cooled enough so as not to react with it. These shields are often used when welding in a gas-purged chamber is not possible. See Figure 6-16.

Figure 6-16 Trailing shield.

What is the GTAW setup process?

Process Setup

- Locate the welding power supply in a dry area, attach the unit to the power source, and set the power switch to OFF.
- Secure the inert gas cylinder with a safety chain so it will not tip over, remove the cylinder cap, crack the cylinder valve to remove any dirt, and using a wrench, attach the regulator to the cylinder. If there is a flow meter tube, make sure it is vertical.
- Connect the gas hoses:
  - From the flow meter or flow regulator to the welding machine

Figure 6-17 A welder prepares the equipment. Notice the work lead clamped to the welding table.

  - Between the welding machine and the torch hose
- If water-cooled, connect the water lines:
  - From the water source to the water IN connection from the water OUT connection on the torch cable to the water drain
  - In the case of a recirculating water cooler, the IN and OUT water lines from the cooler are connected to the torch cable
- Complete the electrical power and control connections between the torch and the welding power supply.
- Insert the electrode in the torch collet and tighten. Install the cup if needed.

- With the electrode and torch positioned so as not to strike an arc, turn on the power supply and apply power to the torch. Check for inert gas and water leaks and tighten fittings if needed, then adjust the gas for proper flow rate.
- Set arc current level, polarity, and secure the work lead to the work.
- Set any other parameters on the welding power supply; these could include:
  - Mode switch if combination GTAW/SMAW power supply
  - High frequency switch: Set to START on DC, CONTINUOUS for AC, and OFF, if pulse waveforms are used
  - Preflow and postflow gas times
  - If using a pulse waveform machine:
    - Ramp-up time (start time)
    - Ramp-down time (stop time)
    - Pulse rate
    - Background current
    - % power-on time
- Select the weld filler metal, if used.

135

Variables Affecting Weld Penetration

- Arc polarity
- Arc current
- Shielding gas

- Size of part being welded
- Thermal conductivity of part
- Preheat temperature, if any

- Melting point of part's metal
- Travel speed

Figure 6-18
Many GTAW jobs call for the use of a filler material. *Photo courtesy of Lincoln Electric.*

Adjusting Shielding Gas Adjust gas flow according to either AWS specification or manufacturer's data sheet. If neither is available, begin at about 20 ft^3/hour (9.5 l/min) and increase gas flow until visible signs of weld porosity cease. Getting an effective inert gas shield from the atmosphere depends on both having enough flow and maintaining a *laminar* (turbulence free) flow. Turbulence will bring air into the area we want to shield, ruining the weld.

• Put on your helmet over your safety glasses, gloves, and you are ready to begin welding. Scratch the tip against the metal and then immediately raise it slightly to produce the arc. metal thicknesses are

Joint Preparation

• Single-pass welds can be made with no preparation on material with a thickness of 0.005 to 0.125 inches (0.13 to 3.2 mm)
• Single-pass welds can be made with preparation on material with a thickness of 0.062 to 3/16 inches (1.6 to 4.8 mm)
• Multi-pass welds can be made with preparations on material with a thickness of 0.125 to 2 inches (3.2 to 51 mm)

Figure 6-19 Consult the manufacturer's data sheet to adjust the shielding gas. The goal is to achieve enough flow and a turbulence-free flow.

Photo courtesy of Lincoln Electric.

What are some problems and solutions for TIG welding

Table 6-2 TIG troubleshooting

Problem	Cause	Solution
Excessive electrode consumption	1. Inadequate gas flow. 2. Operating on reverse polarity. 3. Improper size electrode for current needed. 4. Excessive heating in holder. 5. Contaminated electrode. 6. Electrode oxidation during cooling. 7. Using gas containing carbon dioxide or oxygen.	1. Increase gas flow. 2. Use larger electrode or change to DCEN. 3. Use larger electrode. 4. Check for proper collet contact. 5. Remove contaminated portion. 6. Keep gas flowing after arc stops for at least 10 to 15 seconds. 7. Change to proper gas.
Erratic arc	1. Base metal is dirty or greasy. 2. Joint too narrow. 3. Electrode is contaminated. 4. Arc too long.	1. Use chemical cleaners, wire brush for abrasives. 2. Open joint groove, bring electrode closer to work, decrease voltage. 3. Remove contaminated portion of electrode. 4. Bring holder closer to work.
Porosity	1. Entrapped gas impurities (hydrogen, nitrogen, air, and water vapor). 2. Defective gas hose or loose connections. 3. Oil film on base metal. 4. Too windy.	1. Blow out air by purging gas line before striking arc. 2. Check hoses and connections for leaks. 3. Clean with chemical cleaner not prone to break up arc. 4. Shield work from wind.
Tungsten contamination of workpiece	1. Contact starting with electrode. 2. Electrode melting and alloying with base metal. 3. Touching tungsten to molten pool.	1.Use high-frequency starter, use copper striker plate. 2. Use less current, or larger electrode; use thoriated or zirconium-tungsten electrode. 3. Keep tungsten out of molten pool.

137

TIG Welding Safety

In addition to general safety requirements presented throughout this book, the TIG process requires the following safety measures:

- Because the TIG process is smokeless and there are no visible fumes to block or absorb radiation, heat and light from its arc are intense. Proper lens shade selection is important. Also cover up all skin to prevent radiation burns.

Table 6-3 Lens shade selection chart for TIG

Welding Current (A)	Minimum Protective Shade	Suggested Shade for Comfort
>20	6	6-8
20-100	8	10
100-400	10	12
400-800	11	14

138

- When welding in confined spaces such as inside tanks, provisions must be made for fresh air to replace the inert gas used in the process.
- Good ventilation is vital since the arc generates dangerous ozone levels which are invisible and whose concentrations should be minimized.
- Good ventilation will also prevent argon asphyxiation from slow leaks since argon is heavier than air and will tend to collect in low spots like holes and pits.
- Always be sure to turn off the power supply when changing electrodes as some power supplies have open-circuit voltages as high as 85 volts, enough to be a lethal shock hazard.
- Wear dry gloves without holes; never weld barehanded. Should you touch the filler wire or rod to the electrode, the full voltage of the power supply will be on the filler and you could become part of a lethal circuit. This is especially true if you have worked up a sweat and are sitting on a metal beam or working on a metal table.

TIG welding is one of the more difficult welding skills to master. You must control the torch with one hand, the filler rod with the other and the power feeding the arc using a foot pedal. In addition, the electrode should never touch the work while welding. If it does, you will need to resharpen the electrode. It is difficult, but once you master TIG welding, you can weld pretty much any metal with pinpoint accuracy.

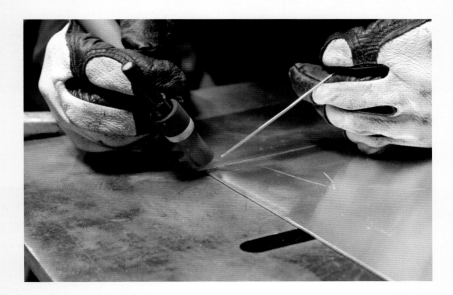

139

Figures 6-20 and 6-21 This welder is tacking welding to keep the work in alignment.

Figure 6-22 Maintaining a consistent travel speed helps ensure a uniform bead along the weld joint.

Figure 6-23 TIG can weld both thick and thin sections of metal. An inert shielding gas keeps the weld free of contamination.

Figure 6-24 Here is an example of a partially welded corner joint. The entire weld area should be fused together.

Thermal Cutting

Oxy-Fuel Cutting and Plasma Arc Cutting

As with welding, heat produced by a combining oxygen and a fuel gas or produced by an electric arc can cut through metal. In fact, much more acetylene is used for cutting metals than for welding them. For many cutting applications, there is no more effective and efficient process. Used in construction, manufacturing, and repair operations, cutting equipment is inexpensive, portable, and easy to use. This chapter explains how the oxygen-fuel cutting process works, its capabilities and limitations. It also covers plasma arc cutting.

How does oxyacetylene cutting work?

How does a cutting head differ

from a welding torch?

What other fuels can be used in oxy-fuel cutting?

How do you set up equipment for for safe operation?

What are some basic cutting techniques?

What are OFC troubleshooting techniques?

What is plasma arc cutting?

Plasma arc cutting tips from West Coast Customs

How does oxyacetylene cutting work?

The process for oxyacetylene cutting (OAC) is similar to that of oxyacetylene welding (OAW). The oxyacetylene flame brings the steel at the beginning of the cut up to kindling temperature of 1,600°F (871°C). At the kindling temperature, steel will readily burn in the presence of oxygen. When the oxygen lever is turned on, the pure oxygen stream along with the steel at kindling temperature causes a chemical reaction called oxidation, which produces slag. This slag has a melting point much lower than the melting point of steel itself and readily runs out of the cut or kerf. The force of the oxygen stream provides additional help to clear the kerf of molten oxides. In addition to the oxyacetylene preheat flame, the burning of the iron in the oxygen stream releases large amounts of heat. This aids cutting action particularly when cutting thick steel. Moving the torch across the work produces continuous cutting action; straight, curved, or beveled cuts are readily made. Conventional OAW equipment (outfit) is readily converted to perform light to heavy OAC by exchanging the welding nozzle on the torch handle to a cutting accessory head fitting into the handle, Figure 7–1.

Steel Thickness Limits. OAC has no practical limit. Steel seven feet thick is routinely cut in heavy industry, and fourteen-foot cuts are not uncommon.

OAC's lower limit is 20 gauge (0.035 inch or 0.88 mm) steel. Below this thickness the cut becomes irregular with uncontrollable melting, but it can be cut with a large tip-to-plate angle and fast travel speed. Thinner steel sheets are best cut with laser or plasma cutters.

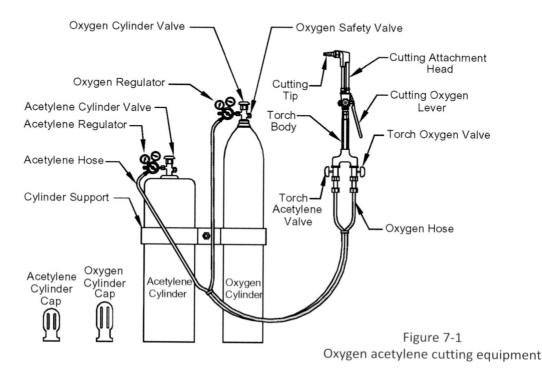

Figure 7-1
Oxygen acetylene cutting equipment

Figure 7-2
This oxyacetylene kit contains
both a welding and cutting accessories

Photo courtesy of Hobart Welders.

Pros and Cons

Advantages

- Low cost compared with machine tool cutting equipment.

- No external power required.

- Readily portable.

- Steels usually cut faster than by conventional machining process.

- Good choice for cutting mild steel, low-alloy steel, cast iron, and titanium

- Cutting direction may be changed easily.

- OAC is an economical method of plate edge preparation for groove and bevel weld joints.

- Large plates may be cut in place.

- Parts with unusual shapes and thickness variations hard to produce with conventional machinery are easily produced with OAC.

- Can be automated using tracks, patterns, or computers to guide the torch.

Disadvantages

- Dimensional tolerance of OAC is dramatically poorer than machine tool based cutting.

- OAC process is limited to steel and cast steel. Definitely not a good choice for aluminum, brass, copper, lead, magnesium, stainless steel, or zinc.

- Both the preheat flame and the stream of molten slag present fire and burn hazards.

- Proper fume control is required.

- Some steels may need pre-heat, post-heat, or both to control the metallurgy and properties of the steel adjacent to the cut.

- High-alloy steels and cast iron need additional process modifications.

How does a cutting head differ from a welding torch?

The OAC cutting head still contains a means of mixing oxygen and acetylene to produce an approximate temperature of 6,300°F (3100°C). But it has added means to deliver a stream of pure oxygen to the cutting point. An oxygen lever opens this pure oxygen stream when the welder depresses it.

The torch on the left side of Figure 7–3 uses an injection chamber or venturi to draw the fuel gas into the oxygen stream and operates with fuel pressures 6 oz/in², such as those supplied by an acetylene generator or a regulated natural gas system delivering in water column inches or about 1/3 pound. The torch depicted on the right uses a mixing chamber to bring the gases together and is also known as balanced-pressure, positive-pressure, or medium-pressure torch. The advantage of the mixing chamber design torch is that it operates at higher fuel gas pressures and can supply more heat than the venturi design: the venturi design when adjusted properly creates a near perfect cutting flame that uses the fuel gas more efficiently. See detail in Figure 7-3.

144

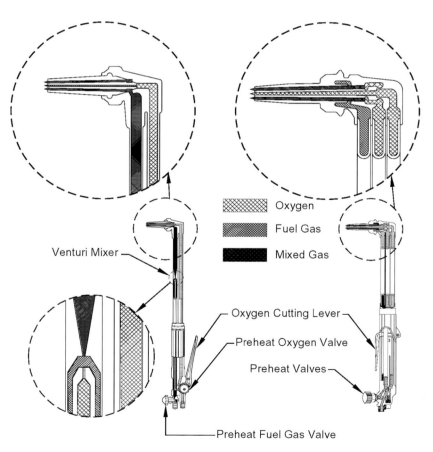

Figure 7-3 Oxy-fuel cutting torches. An injector cutting torch (left) and a mixing chamber positive pressure cutting torch (right)

TOOL TIP

Cutting Tips

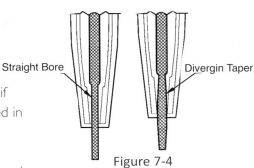

Straight Bore | Divergin Taper

Figure 7-4
Standard cutting tip (left) and high-speed cutting tip (right)

OAC tips are made of copper and can easily be damaged if dropped. Tips from one torch maker cannot, in general, be used in

another manufacturer's torch.

If you have removed the tip nut that retains the torch tip, and the   torch tip is stuck in the torch body, a gentle tap on the back of the torch head with a plastic hammer will release the tip.

Care should be taken when cleaning the tip to avoid breaking

Regular cutting tips have a straight-bore oxygen channel and operate from 30 to 60 psi (2 to 4 bar). High-speed tips have a diverging taper and permit operation at oxygen pressures from 60 to 100 psi (4 to 7 bar). This permits a 20% increase in cutting speed. They are used only on cutting machines. See Figure 7-4.

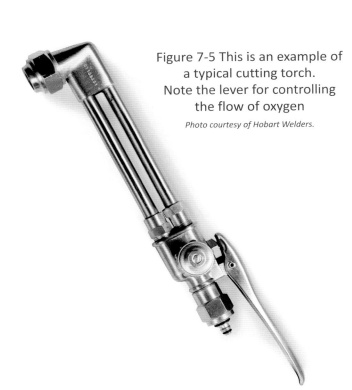

Figure 7-5 This is an example of a typical cutting torch. Note the lever for controlling the flow of oxygen

Photo courtesy of Hobart Welders.

Figure 7-6 The copper tip is used for acetylene; the silver tip is used for propane

Photo courtesy of Hobart Welders.

Other Types of OAC Tips

A wide variety of tips are available. See Figure7-7.

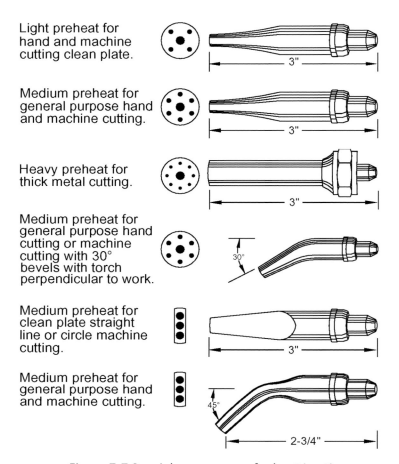

Light preheat for hand and machine cutting clean plate.

Medium preheat for general purpose hand and machine cutting.

Heavy preheat for thick metal cutting.

Medium preheat for general purpose hand cutting or machine cutting with 30° bevels with torch perpendicular to work.

Medium preheat for clean plate straight line or circle machine cutting.

Medium preheat for general purpose hand and machine cutting.

Figure 7-7 Special purpose oxy-fuel cutting tips

What other fuels can be used in oxy-fuel cutting?

When a fuel other than acetylene is used, the process is called oxy-fuel cutting (OFC). These alternative fuels are

- Propane
- Natural gas
- Methyl acetylene-propadiene stabilized (known as MPS or MAPP®)

Although acetylene produces higher pre-heat temperature, alternative fuels are far more stable than acetylene therefore much safer to handle. An alternative fuel may also offer significant cost savings. Fuel selection is a complex matter involving material thickness, cutting speeds, preheat time, fuel performance on straight lines, curves and bevels, and their impact on the total cost. While fuels other than acetylene do not produce the high flame temperature of acetylene, some can produce a greater volume of heat output throughout the outer flame envelope. This gives an advantage to some alternative fuels in cutting thick steels.

In OFC, torch tip designs are frequently different because alternative fuels may be supplied at lower pressures, have different ratios of fuel to oxygen, and different flame and burn rate characteristics. Several manufacturers offer alternative fuel torches. Figures 7-8 and 7-9 shows alternative fuels cutting tips.

Figure 7-8 Alternative fuel gas cutting tip

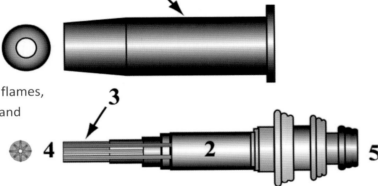

Figure 7-9 Alternative fuel gas tips are two piece tips consisting of:

1 an outer shell,
2 an inner member,
3 grooves for preheating flames,
4 extremities of grooves and
5 a cutting oxygen bore

147

Figure 7-10
Be prepared to make adjustments to the gas flow to achieve and maintain the proper cutting temperatures

How do you set up cutting torch equipment?

- Inspect and clean the torch using the cleaning kit.

- Put on your welding safety equipment: goggles with filter lens (or tinted face shield), cap, high-top shoes, fire retardant coat, cape sleeves and bib or cotton or wool long-sleeved shirt, and pants and welding gloves.

- Avoid wearing trousers with cuffs when cutting as they tend to catch hot sparks and can easily result in a fire. *Wear no synthetics.* If you will be doing overhead cutting, leather skins, fire retardant coats, cape sleeves and bib or aprons are necessary to protect your clothing from falling sparks. Goggles and face shields should have a number 5 shade.

- Firmly secure the oxygen and acetylene cylinders to a welding cart, building column, or other solid anchor to prevent tipping during storage or use. Non-flammable material must be used to secure the cylinders. Remove the safety caps.

- Verify the cutting torch has flashback arrestors installed.

- Check to make sure there are no nearby sources of ignition. Momentarily open each cylinder's valve to the atmosphere and re-close the valve quickly to purge the valve; this is known as *cracking* a valve. Cracking serves to blow out dust and grit from the valve port and to prevent debris from entering the regulators and torch. Stand on the opposite side of the cylinder from the valve port when cracking.

148

- With a clean, oil-free cloth, wipe the valve-to-regulator fittings on both cylinders to remove dirt and grit from the fittings' connection faces and threads. Cleanse both regulators' threads and faces. Remember, never use any oil on high-pressure gas fittings. Oxygen at high pressures can accelerate combustion of oil into an explosion.

- Check to see that both the oxygen and acetylene regulator pressure adjustment screws are loosened (but not falling out of their threads), then screw each regulator to its respective cylinders. Snug up the connections with a wrench. **Caution:** Oxygen cylinder-to-regulator threads are *right-handed*; so are oxygen hose-to-torch screw fittings. Acetylene cylinder-to-regulator fittings and acetylene hose-to-torch fittings are *left-handed* threads. This arrangement prevents putting the wrong gas into a regulator or torch connection.

- Stand so the cylinders are between you and the regulators. S-L-O-W-L-Y open the oxygen cylinders valves. Be sure to open the oxygen cylinder valve until it hits the upper valve stop and will turn no further.

- With the cylinders between you and the regulators, open the acetylene cylinder valve gradually and not more than one and a half turns. If there is an old style removable wrench on the cylinders, make sure to keep it handy in case you must close the cylinder valve immediately in an emergency.

- Look for the high-pressure—or cylinders side—gauges to indicate about 225 psi (15.5 bar) in the acetylene cylinders and 2,250 psi (155 bar) on the oxygen cylinders. These pressures at 70°F (21°C) will indicate the cylinders are fully charged. Note that these pressures will vary with ambient temperature of the cylinders. The pressures given above are for full cylinders at 70°F (21°C), but the actual pressure will vary with cylinder temperature.

- Install the cutting torch on the hoses, or if using a combination welding and cutting handle, install the cutting accessory on the torch handle.

- First, check the area for ignition sources, other than your torch igniter. Then purge each torch hose of air separately: Open the oxygen valve on the torch about three-quarters of a turn, then screw in the pressure control screw on the oxygen regulator to your initial pressure setting. After several seconds, close the torch valve. Do the same for the acetylene hose. **Comment:** We do this for two reasons, (1) to make sure we are lighting the torch on just oxygen and acetylene, not air, and (2) to get the regulators set for the correct pressure while the gas is flowing through them. If the gas hoses are more than 50 feet (15 m) long, a higher regulator setting will be needed to compensate for the pressure drop in the hoses.

- Test the system for leaks at the cylinder-to-regulator fittings and all hose fittings with soapy water. Bubbles indicate leaks.

- Proceed to light and adjust the cutting torch as detailed below.

149

Lighting and Adjusting the Cutting Torch

- Follow the steps of securing the cylinders, installing the regulators, hoses, and torch, purging the hoses of air, and setting the regulator pressures from cutting reference tables for 1/2 inch steel: acetylene at 6 psi (0.4 bar) and oxygen at 30psi (2 bar).

- *Never* adjust the acetylene regulator pressure above 15 psi (1 bar) as an explosive disassociation of the acetylene could occur.

- Open the oxygen valve on the back end of the torch all the way.

- Recheck the low-pressure gauge pressures to make sure the working pressures are not rising. If the working pressure should rise, it means that the regulator is leaking. The cylinders must be immediately shut down at the cylinder valves as continued leaking could lead to regulator diaphragm rupture and a serious accident.

- Light the torch by opening the acetylene valve on the torch handle about 1/16 turn and light the acetylene using your flint igniter. A large, smoky, orange flame will result. Also, you must have your tinted welding facemask over safety glasses (or your welding goggles with a number 5 lens shade) on prior to lighting the flame.

- Increase the flame size by slowly opening the acetylene valve until most of the smoke disappears.

- Open the oxygen preheat valve on middle of the torch and adjust for a neutral flame.
- Actuate the cutting oxygen lever and examine the preheat flame. Further adjustment of the preheat oxygen valve may be needed to keep the preheat flame large enough when the cutting oxygen is used. This is because cutting oxygen use may cause the hose pressure to drop so much the oxygen to the preheat flame must be increased to keep a proper preheat flame.
- You are ready to begin cutting.

Selecting and Adjusting

Given a material and thickness, use a torch manufacturer's table to convert metal thickness to tip size, starting oxygen pressure and acetylene pressure. Remember these are suggested starting pressure ranges. Fine-tuning of the pressures may be needed to get the best combination of speed and quality.

Setup for Alternative Fuels

With a carburizing flames established, adjust the acetylene tips by opening the fuel valve until there is no soot visible at the end of the acetylene flame; then adjust the flame to neutral by adding oxygen. Once the heat cone flames are at neutral, depress the oxygen lever and look at the flame. If it appears to have a feather or carburizing flame, adjust the oxygen with the oxygen stream lever depressed until the flame again looks neutral; now cutting may proceed.

Alternative fuel tips are adjusted by opening the fuel valve enough so that the gas can be ignited followed by adding a small amount of oxygen, reducing the flame enough to see the inner-cones; then alternate between opening the fuel valve a small portion at a time followed by opening the oxygen until you see the heat cones sticking out of the end of the tip approximately 3/16 inch; then increase the oxygen flow until you hear the tip whistle and see the skirt.

Figure 7-11 Appearance of the OFC flame inner-cones and skirt

Shutting Down Oxyacetylene Cutting Equipment

- Turn off the oxygen and then the acetylene with the torch handle valves.

- Turn off the oxygen and acetylene cylinders valves on the cylinders.

- One at a time, open and reclose the oxygen and acetylene valves on the torch handle to bleed the remaining gas in the lines and regulator to the atmosphere. Verify that both the high-pressure and low-pressure gauges on both gases indicate zero pressure. Bleed off the oxygen first to eliminate the possibility of providing oxygen to the remaining acetylene.

- Unscrew the regulator pressure adjustment screws on both regulators in preparation for the next use of the equipment.

What are some basic cutting techniques?

When cutting is performed material is removed, the width of the cut is the *kerf:* when flame cutting, the oxidation of the metal along the line of the cut removes a thin strip of metal, or *kerf,* which is the thickness of the cut, which is also the bore size of the cutting tip. In steel under two inches in thickness, it is possible to hold the kerf to about 1/64 inch (0.4 mm). Cutting thicker steel requires more oxygen, which requires a larger oxygen orifice size, greater oxygen flow rates and a larger oxygen stream. These lead to a wider kerf. See Figure7-12.

151

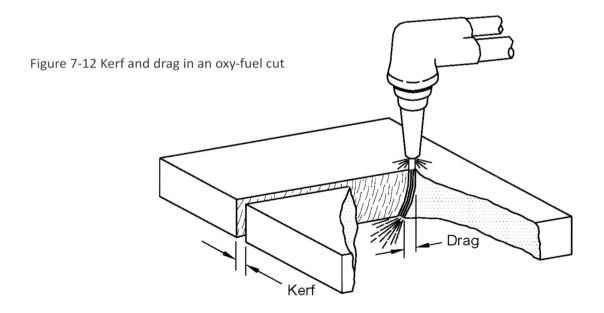

Figure 7-12 Kerf and drag in an oxy-fuel cut

Dealing with Drag

The distance between the cutting action at the top and bottom of the kerf is called *drag*. When the oxygen stream enters the top of the kerf and exits the bottom of the kerf directly below, the drag is said to be zero. If the cutting speed is increased (or the oxygen flow decreased), oxygen in the lower portion of the kerf decreases and the kinetic energy of the oxygen stream drops, slowing cutting action in the bottom of the cut. This causes the cutting action at the bottom of the kerf to lag behind the cutting action at the top. Drag may also be expressed as a percentage of the thickness of the cut. Excessive drag can cause loss of cutting action in thick cuts and restarting the cutting action may cause the loss of a part being flame cut. Excessive oxygen flow, too slow a cut, or damaged orifices may cause reverse drag leading to rough cut edges and excessive slag adhesion.

Selecting Cutting Tips

The numbering system for cutting tip sizing, like welding tip sizing, is not standardized in the welding industry. The drill sizing is standard but the manufacturer's numbers placed on the tips are not standard. In Table 7-1, the cutting drill bore size indicates the cutting orifice size and the thickness o the material that the tip can cut.

Table 7-1 Material thickness to bore size for cutting tips.

Bore Size for Oxy-Fuel Cutting	
Plate Thickness inches (mm)	Bore Drill Size inches (mm)
1/4-1/2 (6.35–12.7)	68-53 DR 0.031-0.059 (0.794-1.51)
3/4 (19.05)	62-53 DR 0.038-0.059 (0.965-1.51)
1 (25.1)	56-53 DR 0.046-0.059 (1.18-1.51)
1 ½-2 (38.1-50.8	51-46 DR 0.067-0.081 (1.70-2.06)
3-5 (76.2-127.0)	46-44 DR 0.081-0.086 (2.06-2.18)
6-8 (152.4-203.2)	40-39 DR 0.098-0.010 (2.49-2.53)
10 (254)	39-35 DR 0.010-0.011 (2.53-2.94)

Cutting ¼- Inch through ½-inch-thick metal

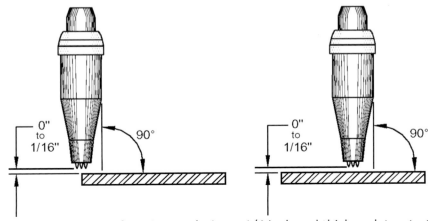

Figure 7-13 Position of cutting torch tip on 1/4 inch and thicker plate, starting (left) and cutting (right)

- Hold the cutting torch tip perpendicular to the metal.

- Start the cut at the edge of the stock by preheating the edge of the stock. In thicker material, the torch may be angled away from the direction of travel so the preheat flame strikes down the edge of the material. When the stock becomes a dull cherry red, begin cutting by squeezing the oxygen cutting lever. Remember to hold the torch tip perpendicular to the surface of the stock when cutting action has begun.

- Move the torch along the cut line in a steady motion. For right-handed welders, cutting from right to left allows the welder to see the marks of the cutting line more easily. Left-handers will usually prefer cutting left to right. See Figure 7-13.

Cutting 1/8-Inch or Thinner Metal

- Utilize the smallest cutting tip available with two preheat flames.

- Hold the torch at a 20-degree to 40-degree angle to the metal surface to increase the kerf thickness.

- Adjust the flame to the smallest preheat flame that will permit cutting.

- Set oxygen pressure at 15psi (1 bar). See Figure 7-14.

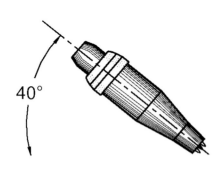

Figure 7-14 Torch position for cutting thin sheet metal, starting (top) and cutting (bottom)

Preventing Slag Accumulation

When cutting thin-gauge sheet metal, if slag accumulates on the underside of the good part, tip the torch away from the side you will use. This allows the slag to form on the scrap side of the kerf keeping it off the good section. Too much slag may indicate that the equipment is not set up properly. Too much slag may indicate that the equipment is not set up properly.

What changes are needed to cut steel thicker than one inch?

Because cutting thick steel requires more oxygen than thin steel, a special oxygen regulator with the capacity of delivering more oxygen volume at higher than welding pressures may be needed. Larger diameter hoses may also be required. The welding acetylene regulator is fine for cutting. Also since there is much higher oxygen consumption and more rapid cylinder depletion than in welding operations, the typical cutting regulator is a two-stage regulator to maintain a constant working pressure as the cylinder gas dwindles. Oxygen regulators specifically for cutting usually have low-pressure gauges (on the output or torch side of the regulators) with higher pressure calibrations than welding regulators. OFC operations on extremely thick metals can require 100 to 150 psi (6.8 to 10 bar) oxygen pressures.

Table 7-2 Optimum pressure and gas flow settings for cutting various metal thicknesses

Plate Thickness in inches	Oxygen PSIG	Fuel Gas PSIG	Oxygen bore drill size
1/8	20-25	3-5	0.031
1/8-1/4	20-25	3-5	0.036
¼-1/2	25-35	3-5	0.040
½-3/4	30-35	3-5	0.046
3/4-1 1/2	35-45	3-7	0.059
1 ½-2/12	40-50	4-10	0.067
2 1/12-3	45-55	5-10	0.093
3 1/2-5	45-55	5-10	0.110
5-8	45-60	7-10	0.120

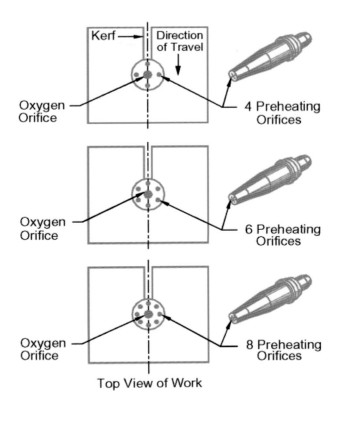

Figure 7-15 Location of preheat orifices in relation to kerf for a normal cut

Adjusting Torch Tips

If there are two preheat orifices, the tip should be rotated in the torch so that a line drawn between orifices will be perpendicular to the cut line. If more orifices, two should fall on the cut line and the rest divided equally on each side of the cut line. This symmetrical preheating improves the quality of the cut. See Figure 7-15 for normal cut. For making a bevel cut, see See Figure 7-16.

155

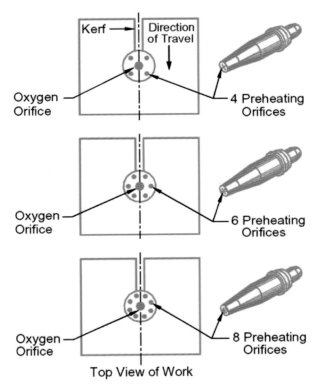

Figure 7-16 Location of preheat orifices in relation to kerf for a bevel cut

Starting a Heavy Cut

See Figure 7-17.

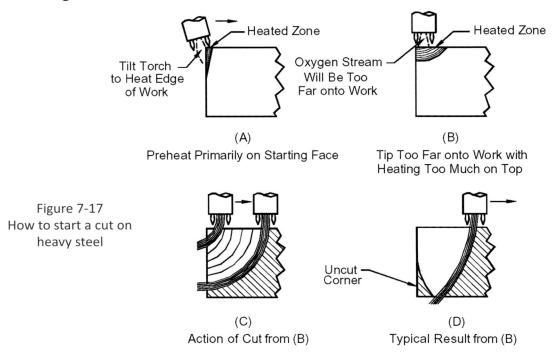

Figure 7-17
How to start a cut on
heavy steel

(A)
Preheat Primarily on Starting Face

(B)
Tip Too Far onto Work with
Heating Too Much on Top

(C)
Action of Cut from (B)

(D)
Typical Result from (B)

Ending a Heavy Cut

As the end of the cut nears, tilt the torch away from the direction of travel. This permits the bottom of the cutting action to proceed ahead of the top cutting action and eliminates premature breakout of the flame, which leaves a triangle at the end of the cut. See Figure 7-18.

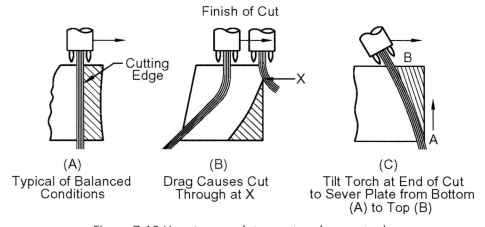

Finish of Cut

(A)
Typical of Balanced
Conditions

(B)
Drag Causes Cut
Through at X

(C)
Tilt Torch at End of Cut
to Sever Plate from Bottom
(A) to Top (B)

Figure 7-18 How to complete a cut on heavy steel

Piercing Steel with OFC

Begin by preheating the material in the pierce location to a dull red color, kindling temperature, with the torch perpendicular to the metal. When metal becomes dull red, slightly raise the torch from the surface and angle the tip away from perpendicular. This prevents the slag blown back from the surface from landing on or in the torch tip. Then squeeze the oxygen lever to start cutting action. As soon as the material is completely pierced, restore the tip to perpendicular and the preheat flame to just above the surface. Complete cutting the opening wanted.

If a small hole is wanted and the surrounding material is to be protected from cutting action, drill a 1/4 inch (6 mm) hole at the starting point. Begin the cutting action through the hole. See Figure 7-19.

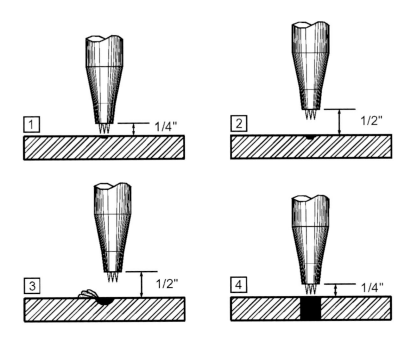

Figure 7-19 Piercing steel

Cutting Out a Circle

Pierce the material inside the circle and away from the finished edge. When cutting action is established, extend the cut into a spiral and begin cutting the circle itself, Figure 7-20. With small circles to avoid damaging the finished edge, drill a 1/4 inch (6 mm) hole in the center of the circle and begin the cut through the inside of the hole, then spiral out to the edge.

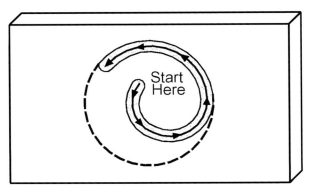

Figure 7-20 Cutting a circle

What are OFC troubleshooting techniques?

Bell-mouth kerf is caused by excessive oxygen pressure, see Figure 7-21.

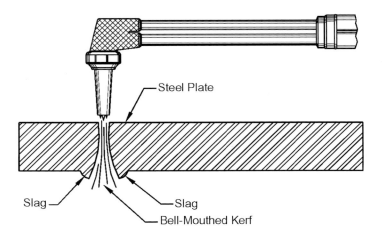

Figure 7-21 Bell-mouthed cut

Correcting Defective Edges

Compare the defective edge with the drawings in Figure 7-22 to diagnose the problem.

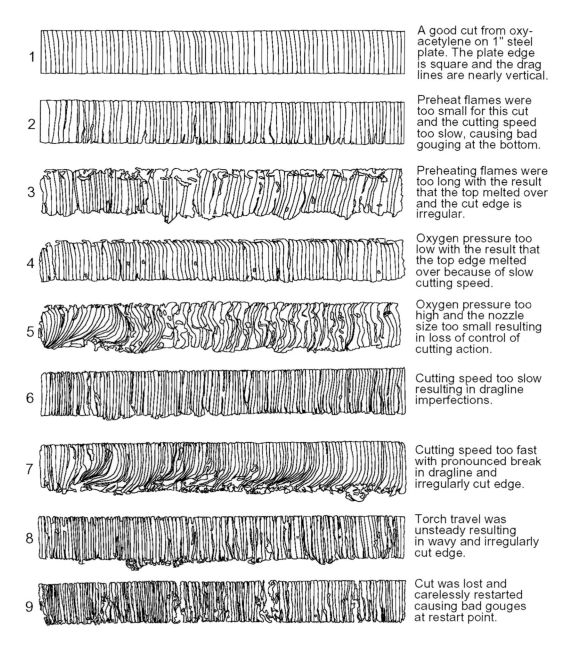

1. A good cut from oxy-acetylene on 1" steel plate. The plate edge is square and the drag lines are nearly vertical.

2. Preheat flames were too small for this cut and the cutting speed too slow, causing bad gouging at the bottom.

3. Preheating flames were too long with the result that the top melted over and the cut edge is irregular.

4. Oxygen pressure too low with the result that the top edge melted over because of slow cutting speed.

5. Oxygen pressure too high and the nozzle size too small resulting in loss of control of cutting action.

6. Cutting speed too slow resulting in dragline imperfections.

7. Cutting speed too fast with pronounced break in dragline and irregularly cut edge.

8. Torch travel was unsteady resulting in wavy and irregularly cut edge.

9. Cut was lost and carelessly restarted causing bad gouges at restart point.

Figure 7-22 Cutting Problems and their causes

What is plasma arc cutting?

Plasma arc cutting, AWS designation *PAC*, is an arc cutting process that uses a constricted arc and removes molten metal with a high-velocity jet of ionized gas issuing from a constricting orifice. There are two variations:

1 The first variations are in low-current plasma systems which use the nitrogen in compressed air for the plasma and are usually manual.

2 The second variations are in high-current plasma systems which use pure nitrogen for the plasma and are usually automatic.

The PAC torch works very much like the plasma arc welding torch. Plasma heat input is very high and melts the work metal. Then the plasma jet blows away the molten material completing the cut. Some PAC systems inject water into the plasma to reduce fumes and smoke; others perform the cutting under water to reduce noise and airborne metal vapor.

Figure 7-23 Plasma arc equipment can cut through any metal

Capabilities of PAC?

High-current PAC systems cut 1/8 inch (3 mm) thick metal with a 100 inch/minute (2.5 m/minute) travel speed, 0.050 inch (1.25 mm) thick metal with a 200 inch/minute (5 m/minute) travel speed. The smaller, hand-held torches are used in sheet metal and auto body work. Attachments are available to convert plasma arc welding torches for PAC.

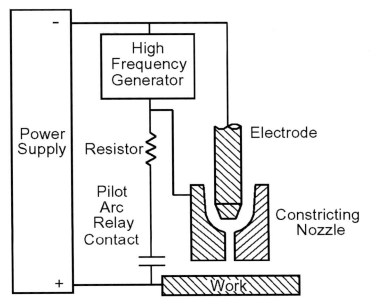

Figure 7-24 PAC schematic

161

Pros and Cons of PAC

Advantages

- PAC cuts all metals.
- Cutting action is so rapid that despite the high heat input, there is a smaller heat affected zone than in most other processes.
- PAC can pierce metals cleanly without the starting hole needed by OAC.
- PAC is ideal for cutting parts under computer-driven control.
- With its 30,000°F (16,600°C) plasma, it cuts materials with melting points too high for OAC.
- All positions can be used.
- Surface smoothness of the cut edges is equal to or better than OAC.

Disadvantages

- Equipment may be expensive. Small units today are more cost effective than in the past.
- Metal vapor produced from the cutting must be captured.
- Thick cuts are normally done under water so the metal vapor can be captured; the water container must then be periodically cleaned usually requiring a HAZ MAT crew.

Rules For PAC Cutting

All of the safety rules suggested in OAC cutting should be applied to PAC including:

- This process uses electricity with voltage ranges from 150 to 400 volts of direct current; this equipment must be properly grounded to avoid electric shock.
- Keep electrical circuits dry.
- Keep all mechanical electrical connections tight; this includes the work lead. Poor electrical connection can cause over heating and fires.
- Proper ventilation is required to prevent inhalation of hazardous metal vapors and gases.
- When securing the equipment, always be sure the power supply has been properly shut down and the torch placed back in its properly insulated storage position.

West Coast Customs TIPS PAC equipment is rated for the thickness of the material it can cut. Many newer machines contain an internal air compressor, which eliminates the need for external air tanks for some cuts. Check the manufacturer's data sheet to determine when you need to add external air. When cutting, be sure to wear the appropriate clothing to protect your skin from sparks and the appropriate shade lens to protect your eyes. Follow the manufacturer's instructions for operating and using the equipment.

Figure 7-25 Here's a good safety feature. The yellow flap is a trigger guard that will keep you from pressing the trigger before you are ready

Figure 7-26 Watch the sparks to determine if your travel speed is correct. Moving too fast will prevent the arc from cutting through the metal, and the sparks will fly back toward you. Move at the right speed and the sparks go to the floor

Figure 7-27 Plasma will cut anything that conducts electricity. Here scrap lumber serves as a straightedge for cutting

Figure 7-28 In addition to straight cuts, PAC allows you to make any shape cut you need, including circles. Any dross along the cut edge can be cleaned up with a small file

Brazing & Soldering

Soldering has been used for thousands of years, but a solid theoretical understanding has come only in the last one hundred. Brazing is similar to soldering but performed at higher temperatures. It has come into use more recently as higher temperature torches became available. Although soldering and brazing do not make joints as strong as welded ones, they are widely used in making and repairing a wide range of products from airplanes to computers to household plumbing to jewelry. We will discuss the theory, materials, fluxes, and common processes. These processes have some advantages over welding and we will present them. There are safety issues, too. Because soldering copper water pipe is so useful and something every welder should know how to do, we cover step-by-step instructions.

What's the difference between brazing and soldering?

Brazing joins materials that have been heated to the brazing temperature followed by adding a brazing filler metal that has a melting point above 840°F (450°C). This temperature will be below the melting point of the base metals joined by this joining process. The non-ferrous filler metal is drawn into and fills the closely fitted mating or *faying* surfaces of the joint by capillary attraction. The filler material, usually aided by fluxes, *wets* the base metal surfaces allowing the brazing material to flow or capillary more readily through and between the two surfaces. When the filler metal cools and solidifies, the base materials are joined. See Figure 8-1.

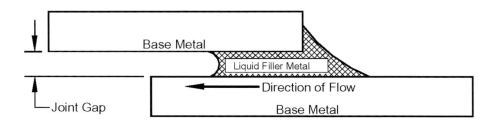

Base Metal

Liquid Filler Metal

Direction of Flow

Joint Gap

Base Metal

166

Figure 8-1 Capillary forces draw molten filler metal into brazed joint

The above is also the definition of soldering, except that brazing takes place *above* 840°F (450°C) and soldering *below* 840°F (450°C), otherwise, the processes are quite similar. They both depend on capillary attraction to draw filler metal or solder into the joint. In general, brazed joints are stronger than soldered ones because of the strength of the alloys used.

The term braze welding is also used. Brazing depends on capillary attraction to draw the filler metal into the mating joint, while in braze welding the filler metal is deposited in grooves or fillets at the points where needed for joint strength. Capillary attraction is not a factor in distributing the filler metal, as the joints are open to the welder. Braze welding is *not* a brazing process, but welding with brazing filler metal. Braze welding is frequently used to repair cracked or broken cast-iron parts. Joint design is similar to those for oxyacetylene welding: V-groove butt joints, lap joints, T-joints, fillets, and plug joints.

Brazing and Soldering

Most common metals can be brazed or soldered including:

- Aluminum
- Bronze
- Brass
- Cast iron
- Copper

- Stainless Steel
- Steel
- Titanium
- Tool steels (some)
- Tungsten carbide (tool bits)

Pros and Cons

Advantages

- Ability to join dissimilar metals—steel is easily joined to copper, cast iron to stainless steel, and brass to aluminum. Many combinations of metals are readily joined.
- Ability to join nonmetals to metals—ceramics are easily joined to metals, or each other.
- Ability to join parts of widely different thicknesses—either thin-to-thin or thin-to-thick parts may be joined without burn-through or overheating.
- Excellent stress distribution—many of the distortion problems of fusion welding are eliminated because of the lower process temperature and the even distribution of heat with more gradual temperature changes.
- Low temperature process—the components being joined are less likely to be damaged because the base metals are not subjected to melting tems peratures.
- Joins precision parts well—with proper jigs and fixtures, parts may be very accurately positioned.
- Parts may be temporarily joined, subjected to other manufacturing processes, and then separated without damage.
- Mistakes are readily fixed—a misaligned part can be repositioned without damage.
- Ability to make leak-proof and vacuum-tight joints
- Joints require little or no finishing—with proper process design the brazed or soldered joint can be nearly invisible.

Disadvantages

- While brazing processes can produce high-strength joints, they are rarely as strong as a fusion-welded joint.
- The brazed parts and the filler metal may lack a color match.

What are the types of brazing/soldering?

Torch Brazing

Heating is done with one or more gas torches. Depending on the size of the parts and the melting point of the filler metal, a variety of torch fuels (acetylene, propane, methylacetylene-propadiene stabilized, or natural gas) may be burned in oxygen, compressed air, or atmospheric air. A neutral or oxidizing flame will usually produce excellent results.

Flux is required with most braze filler metals and can usually be applied to the joint ahead of brazing. The filler metal can be preplaced in the joint or face fed. Manual torch brazing is successfully used on assemblies involving components of unequal mass. It is frequently used in the repair of castings in the field and is often automated for high-production of small and medium-sized parts. It is a versatile process and probably the most popular brazing technique.

Torch Soldering

This is very much like torch brazing but at a lower temperature. Usually propane, methylacetylene-propadiene gas, or natural gas burning in air supplies the heat.

The joint is cleaned to shiny metal with emery cloth, wire brushes, steel wool, or commercial abrasive pads. The flux (used for wetting the joining surfaces) is usually applied in liquid or paste form, or it may be alloyed inside the solder wire.

While widely used in manufacturing and maintenance, it is also used in plumbing to join copper tubing for potable water. See page 179 for a detailed procedure for the torch soldering of copper tubing.

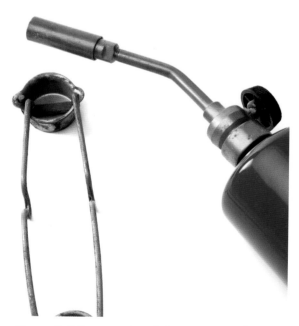

Figure 8-2 Propane torch is a common heat source for soldering.

Iron Soldering

Traditional soldering irons contain a copper tip on a heat-resistant handle. They are heated electrically or in a gas, oil, or coke furnace. The copper tip stores and carries heat to the solder joint. This transfer is made possible by heat being transferred from the heated tip of the iron to the part to be soldered; when the joint is raised to soldering temperature, solder is applied to the joint itself and wets the entire joint. Flux core solder is used for electronic work. This solder has been formed concentrically around a core of one or more strands of flux. In sheet metal and other non-electrical work, the flux may also be in the solder core or applied as a paste or a liquid.

Today most soldering irons are heated electrically and are available from just a few watts for electronic work to 1250 watts for roofing and heavy sheet-metal work. Many irons for electronic work have temperature-controlled tips to avoid damage to the sensitive components. See Figure 8-3

Figure 8-3 An electric soldering iron is used to make repairs to a circuit board

What are some common joint designs?

Well-designed braze joints start with fundamental butt and lap joints. Figures 8-4A shows a butt joint angled to provide more surface area for the brazed or soldered material to bond; this joint is called a scarf joint. Figure 8-4B shows both a lap joint and a square edged butt joint. Good joint design ensures a reliable, repeatable production brazing process that will provide a strong joint.

Joint clearances range from 0.002 to 0.010 inches (0.05 to 0.25 mm). Assuming that the joint clearance is adequate to admit braze filler material, the lower the joint clearance, the stronger the joint. Too much clearance will reduce joint strength, too little will permit voids in the joint.

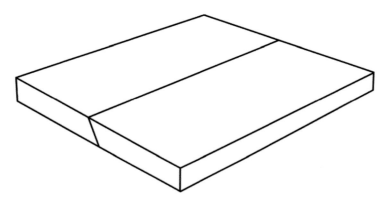

Figure 8-4A Butt joint prepared at an angle and called a scarf joint

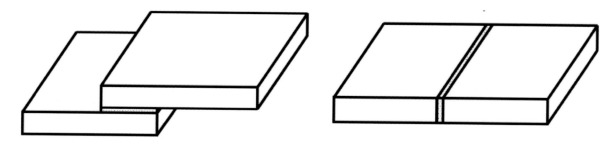

Figure 8-4B Lap and butt joints

Increasing Strength

Design changes to increase butt and lap joint strength include the following: See Figures 8-5 through 8-7.

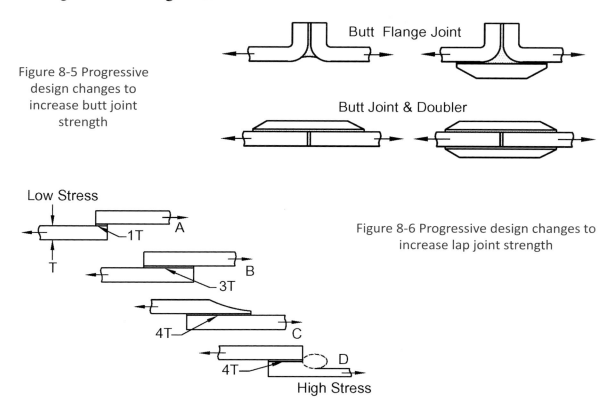

Butt Flange Joint

Figure 8-5 Progressive design changes to increase butt joint strength

Butt Joint & Doubler

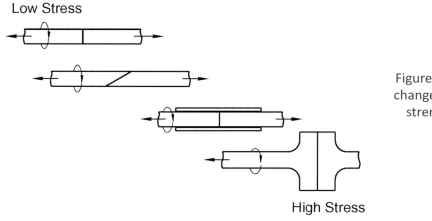

171

Figure 8-6 Progressive design changes to increase lap joint strength

In Figure 8-6, the maximum strength of a simple lap joint appears in B, an overlap of 3 times the thickness of the base metal (3T). More overlap without additional refinements will not improve strength.

Figure 8-7 Progressive design changes to increase butt joint strength against torsional (twisting) forces

Good Joint Design

The factors that influence joint good design include:

- Base metal selection—differences in base metal thermal expansion coefficients may lead to poor joint fit at brazing temperature with too much or too little clearance.
- Effect of flux on clearance—flux must enter the joint *ahead* of the braze filler metal and then be displaced by it, but when joint clearance is too small the flux may be held in the joint by capillary forces. This prevents proper braze filler entry and leaves voids.
- Effect of base metal to filler metal combinations on clearance.
- Effect of brazing metal filler on clearance.
- Effect of joint length and geometry on clearance.
- Dissimilar base metals form a cell that leads to electrolytic corrosion.

Joint Preparation

There are two classes of cleaning: chemical and mechanical. There are many different processes in use. Some common mechanical ones are:
- Grinding
- Filing
- Machining
- Blasting
- Wire brushing

Wire brushes must be free of contaminating materials and selected so none of the wire wheel material is transferred to the part being cleaned. A stainless steel wire brush is a good choice for most materials.

Blasting media must be chosen so it does not embed in the base metal and is easily removed after blasting. For this reason, blasting media like alumina, zirconia, and silicon-carbide should be avoided. These processes are used to remove all dirt, paint and grease so the flux and braze filler metal can readily and completely wet the base metal surface.

Many of the same processes used in brazing are used for soldering. However, in a non-production situation, mechanical cleaning especially with emery cloth, steel wool, or commercial abrasive pads or by filing will be effective. Getting down to fresh, bare metal is the objective. Complete the soldering immediately, before the base metals have a chance to re-oxidize.

Figure 8-8 Copper tubing is cut to length before soldering

173

What's the purpose of brazing and soldering fluxes?

Fluxes do the following:
- Further clean the base metal surface after the initial chemical or mechanical cleaning.
- Prevent the base metal from oxidizing while heating.
- Promote the wetting of the joint material by the braze filler material or solder by lowering surface tension and to aid capillary attraction.

Flux covers and *wets* the base metal preventing oxidization until the braze filler material or solder reaches the joint surfaces. Since the flux has a *lower* attraction to the base metal's atoms than the filler or solder, when the filler metal or solder melts, it slides *under* the flux and adheres to the clean, unoxidized base metal surface ready to receive it. Fluxes will not remove oil, dirt, paint or heavy oxides, so the joint surface must already be clean for them to work.

Brazing Fluxes

Brazing fluxes usually contain fluorides, chlorides, borax, borates, fluoroborates, alkalis, wetting agents, and water. A traditional and still common flux is 75% borax and 25% boric acid (borax plus water) mixed into a paste.

The *AWS Brazing Manual* provides specifications for brazing and brazing fluxes. This specification has 15 classifications of fluxes. Many manufacturers supply proprietary flux mixtures meeting these specifications. See Table 8–1 for abbreviated AWS flux categories and applications.

Table 8–1 Representative brazing flux categories Mg = Magnesium

Base Metals Being Brazed	Filler Metal Type	AWS Flux Class.	Typical Flux Ingredients	Activity Temperature Range		Application
				°F	°C	
All brazeable aluminum alloys.	BAlSi	FB1-A (Powder)	Flourides Chlorides	1080-1140	560-615	For torch or furnace brazing.
All brazeable ferrous and non-ferrous metals except those with Al or Mg as a constituent. Also for brazing carbides.	BAg and CuP	FB3-A (Paste)	Borates Fluorides	1050-1600	565-870	General purpose flux for most ferrous and non-ferrous alloys. Notable exception is aluminum bronze.
Same as above.	BAg, BCuP, and BCuZn	FB3-K (Liquid)	Borates	1400-2200	760-1205	Exclusively used in torch brazing by passing fuel gas through a container of flux. Flux is applied by the flame.
Brazeable base metals containing up to 9% Al (Al-brass, Al-bronze, Monel® K500).	BAg and BCuP	FB4-A (Paste)	Chlorides Fluorides Borates	1100-1600	595-870	General purpose flux for many alloys that form refractory oxides.

Note: The letter *B* in filler metal indicates brazing usage, not the chemical element boron.

Al = Aluminum Si = Silicon Ag = Silver Cu = Copper Mg = Magnesium P = Phosphorous Zn = Zinc

Soldering Fluxes

The main types of soldering fluxes are

Organic fluxes—consisting of organic acids and bases, after soldering they can be removed with water and are widely used in electronics.

Inorganic fluxes—containing no carbon compounds, so they do not char or burn easily and are used in torch, oven, resistance, and induction soldering. They are not used for soldering electrical joints.

Rosin-based fluxes—easily cleaned from parts after soldering. They are usually non-corrosive; available as powders, pastes, liquids, and as a core within soldering wire. They are used in electrical and electronics applications.

Flux Application

For Brazing:

- Apply flux by spraying, brushing, or dipping.
- Heat the brazing rod using a torch and dip the rod into powder flux.
- Some brazing (and braze welding) rods come from the factory with flux already applied to the outside.
- Sheets, rings, and washers of flux can be inserted in the joint before assembly.
- Special guns can inject flux (or mixtures of flux and filler metal) directly into the joint.
- Flux can be dissolved in alcohol and supplied within the fuel gas stream directly to the brazing joint eliminating the manual operation of adding flux. This process automatically controls the amount applied. See Figure 8-9.

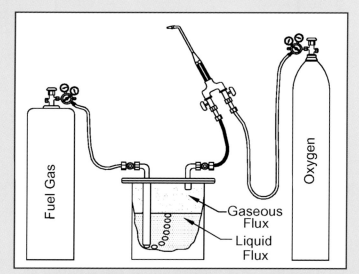

Figure 8-9 Gas fluxing unit for oxy-fuel brazing

For Soldering

Soldering fluxes are brushed, rolled, or sprayed. Many solders have flux cores, so no separate fluxing step is needed.

What are the properties of brazing filler materials and soldering alloys?

Brazing filler materials must have the ability to make joints with mechanical and chemical properties for the application.

- Melting point below that of the base metals being joined and with the right flow properties to wet the base metals and fill joints by capillary attraction. See Figure 8-10.
- Composition that will not allow it to separate into its components during brazing.

176

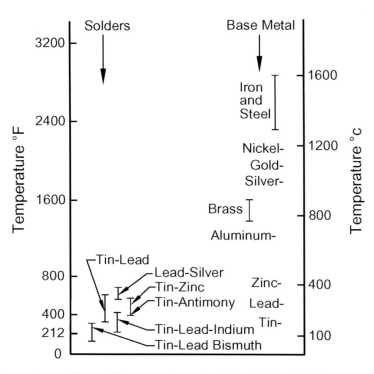

Figure 8-10 Melting points of braze filler metals and solders fall well below most base metals

Common Brazing Fillers

Filler materials covered by AWS specifications are grouped as:

- Aluminum-silicone
- Copper
- Copper-phosphorus
- Copper-zinc

- Heat-resisting material
- Magnesium
- Nickel-gold
- Silver

Figure 8-11 An artist using filler material while working on a sculpture

177

Common Solders

Tin-lead alloys are the most common solders. It is customary to indicate the tin percentage first, then the lead content and the same with other two-metal alloys. A 40/60 tin-lead solder is 40% tin and 60% lead.

Tin-lead solders 35% / 65%, 40% / 60% and 50% / 50% are popular because of their low liquidus temperatures, which is the lowest temperature at which a metal is completely liquid. Tin-lead solders 60% / 40% and the 63% / 37% eutectic are used when low-processing temperatures are needed. Tin-silver, tin-copper-silver, and tin-antimony alloys are used where lead must be eliminated for health reasons as in stainless-steel fabrication for kitchens, food processing equipment, and copper potable water systems. Never use lead-containing solder on potable water systems.

What steps help ensure safe brazing and soldering?

Base metals and filler metals may contain toxic materials, such as antimony, arsenic, barium, beryllium, cadmium, chromium, cobalt, mercury, nickel, selenium, silver, vanadium, or zinc. These will be vaporized during brazing or soldering and cause skin, eye, breathing, or serious nervous system problems. Some of these toxic materials are cumulative such as lead and may be absorbed through the skin. The following precautions are essential:

- Keep your head out of the brazing or soldering plume.
- Perform brazing or soldering in a well ventilated area.
- On failure of normal ventilating equipment, use respiratory equipment.

Many brazing and soldering fluxes and heating bath salts contain fluorides. Others contain acids and aluminum salts. The following precautions apply:

- Avoid direct contact with skin.
- Do not eat or keep food near these materials.
- Do not smoke around these materials. Keep fire extinguishers close at hand.
- Insure MSDSs are affixed to containers of these materials and major equipment using them so they are visible to you and others.
- Protect skin from sparks and hot metal by wearing the appropriate gloves and nonflammable clothing.
- Provide adequate ventilation when using cleaning solvents to prepare the joints; chlorinated hydrocarbons are toxic and may create phosgene gas when heated.

Eye Protection

For soldering, wear safety glasses or face shields to protect the eyes from external injuries caused by sparks, flying metal, or solder splashes.

For brazing, using a number 5 tinted lenses will protect against internal (retinal) eye damage caused by viewing the radiation coming off hot metal. Some brazing requires darker lens shades of up to number 8.

How do you solder copper tubing?

Brazing and soldering have many uses, but the most common is using solder to join sections of copper tubing in home water systems. By U.S. Federal law, only lead-free solders, pipe, and fittings may be sold or used in drinking water systems. These solders are commonly 95-5 tin-antimony, but may include a variety of other alloys of about 95% tin combined with copper, nickel, silver, antimony, bismuth, or other alloying elements. These lead-free solders can cost 4 to 8 times as much as the tin-lead alloys they replace.

Drainage Systems

Lead containing solders may be used in drainage systems. Most common are 50-50 tin-lead and 60-40 tin-lead alloys, because they cost less than lead-free solder alloys. However, with the exception of the 95-5 tin-antimony solder alloy, the lead-free solders have a wider pasty temperature range than the traditional 50-50 and 60-40 tin-lead solders, are easier to handle, and are gaining popularity even where a lead-free solder is not required.

Torch Selection

Air-acetylene, propane, and MAPP® gas torches all work well. The Bernz-O-matic® torches in Figure 8-12 burn either propane or MAPP gas. Changing the gas orifice is all that is required to switch between fuels. MAPP gas is more expensive than propane and has a hotter flame, so it can heat the pipe and fitting faster and handle larger diameter tubing. The version with a separate regulator, 48-inch (1.2 m) hose, and hand piece is especially convenient. It can put heat into tight quarters and does not stall out when inverted, as the single-piece torch does.

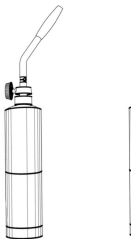

Figure 8-12 Torch-on-cylinder model (left) and hose-based model (right)

Preparing Copper Tubing

- Measure the tubing length for cutting. The tubing must be long enough to reach to the cup (bottom) of its fitting, but not so long as to cause stress in the completed piping.
- Cut the tubing, preferably with a disc-type tubing cutter to insure square ends. A hacksaw, an abrasive wheel, or a portable or stationary band saw may also be used.
- Remove burrs from inside the tubing ends with a round file, half-round file, or reamer. Many disc-type tubing cutters carry a triangular blade for inside reaming. Remove outside burrs with a file because these burrs prevent proper seating of the tubing into the fitting cup. A properly reamed piece of tube provides an undisturbed surface through the entire fitting for smooth, laminar flow, and minimum pressure drop.
- Use emery cloth, nylon abrasive pads like 3M™ Scotch-Brite®, or male and female stainless steel brushes sized to the tubing, to remove the dirt and oxide from the end of the tubing and the inside mating surfaces of the fitting. An electric drill can drive the male stainless steel internal brush and save a lot of elbow grease on a large job. The goal is to get down to fresh, shiny metal. Clean the outside of the tubing from the end of the tubing to about 3/8 inch (1 cm) beyond where the tubing enters the fitting. The capillary space between the tubing and its fitting is about 0.004 inches (0.1 mm). Filler metal fills this gap. Removing too much metal from either the tubing or the fitting will prevent proper flow of solder around the tubing by capillary forces and weaken the joint. Do not touch the cleaned area with bare hands or oily gloves. Skin oils, lubricating oils, and grease impair the adherence of filler metal.

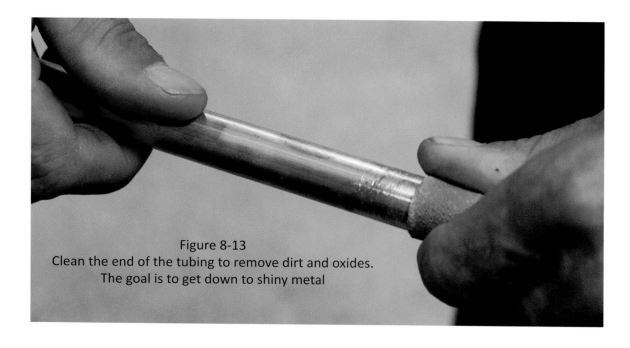

Figure 8-13
Clean the end of the tubing to remove dirt and oxides.
The goal is to get down to shiny metal

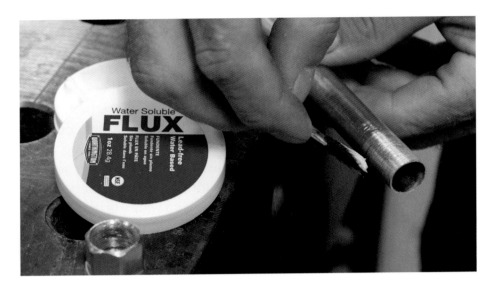

Figure 8-14
Coat the end of the
tubing with flux that is
recommended for the
solder you are using

- Select the flux to match the solder, since a mismatch may cause problems. Use a flux meeting ASTM B 813 requirements.
- Using a brush, apply a thin coating of flux to the ends of the tubing sections and the fitting's interior mating surfaces. See figure 8-14 and 8-15. A disposable acid brush works well.
- Insert the tubing end into the fitting with a twisting motion. Make sure the tubing is properly seated in the fitting cup. With a cotton rag, wipe excess paste flux off the tubing from the exterior of the joint. **Caution:** Do not leave a cleaned and fluxed joint unsoldered overnight. If you must stop work before soldering, disassemble the joint(s) and remove all the flux. On starting work, the next day, perform the cleaning and fluxing process from the beginning.
- Most of the time there are two or three tubing lengths going in a fitting. Insert all the tubing lengths into the fitting so they can all be soldered in one operation. This will save work time and torch fuel. Make sure the tubing lengths and fittings are supported or braced so that all tubing remains at the bottom of its respective fitting cup. You are now ready to solder the fitting.

181

Figure 8-15
Apply flux
to the interior
of the fitting

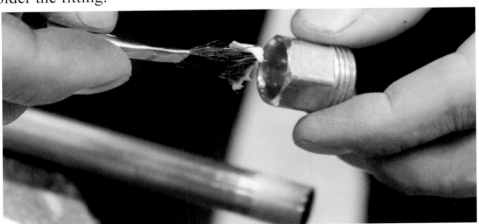

Heating the Tubing

- Begin by heating the copper tubing outside and just beyond the fitting, Figure 8-16. This brings the initial heat into the fitting cup and provides even heat to the joint area. If the joint is vertical be sure to heat the tubing around its entire circumference. If the joint is horizontal, heat the bottom and sides of the tubing. Do not heat the top of the tubing to avoid burning the flux. The natural tendency of heat to rise preheats the top of the assembly and brings sections of tubing up to temperature. The larger the joint, the more heat and time will be needed to accomplish preheating. Experience will determine the amount of preheat time needed. Large-diameter tubing is best soldered with two torches or a multi-orifice torch.

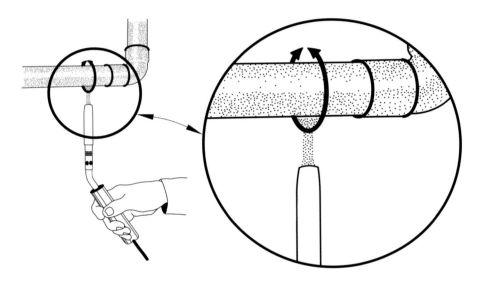

Figure 8-16. Preheat the tubing around the bottom and sides about one cup length away from the fitting

- Now move the flame onto the base of the fitting cup and sweep the flame back and forth between the base of the fitting cup and along the tubing sides to a distance equal to the depth of the fitting cup, Figure 8-17. Preheat the circumference of the assembly as described above. Again, do not heat the top during this preheating to prevent burning the flux.

- With the flame positioned at the base of the fitting cup, touch the solder to the joint at the point where the tubing enters the fitting, Figure 8-18. Keep the torch flame from playing directly on the solder. If the solder does not melt, remove it, and continue preheating. When the solder melts when touching against the fitting and tubing, soldering can begin. **Caution:** Do not overheat the joint or direct the flame onto the face of the fitting cup. Overheating could burn the flux, destroying its effectiveness, and preventing the solder from entering the fitting properly.

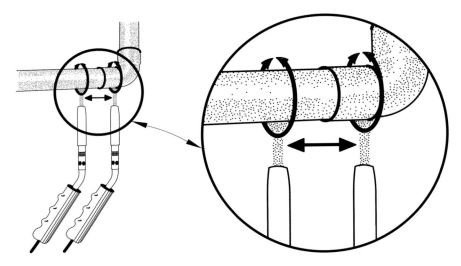

Figure 8-17. Preheat the fitting starting from the base of the cup and sweep out
to the previous preheat zone

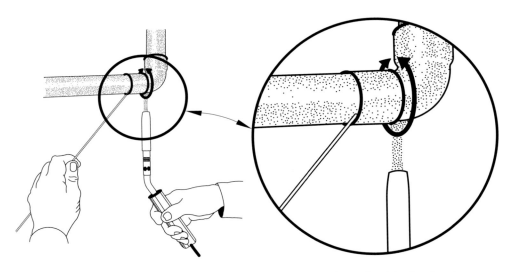

Figure 8-18. Determining when preheating ends and soldering begins

For joints in the horizontal position:

- Start applying the solder slightly off-center at the bottom of the joint, as in Figure 8-19 on the next page. When the solder begins to melt, push the solder into the joint while moving the torch upward along the base of the fitting and slightly ahead of the point of application of the solder. When the bottom portion of the fitting fills with solder, a drip of solder will appear at the bottom of the fitting. Continue to feed solder into the joint while moving the solder up and around to its top or 12 o'clock position, and at the same time move the flame so it slightly leads the solder up to the top of the joint. In general, solder dripping off the fitting bottom is not coming from the bottom of the joint. As the joint fills, excess solder runs down the face of the joint and drips off the bottom. This is a good indication of using the proper amount of solder and a full joint. This will allow the solder in the lower portion of the joint to become pasty, form a dam, and support the solder applied above it, Figure 8-20 (a) and (b).

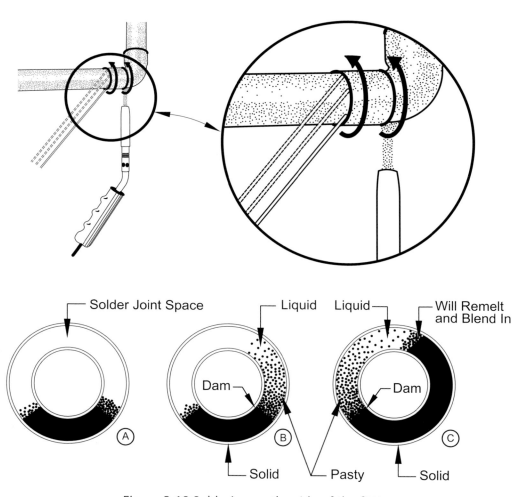

Figure 8-19 Soldering up the side of the fitting
Figure 8-20 Sectional view of a solder fitting showing how cooler solder in the base of fitting forms a dam to hold pasty solder in place until it cools

• Move torch heat and solder to the other side of the fitting and overlapping slightly, proceed up the uncompleted side to the top and over, again over-lapping slightly, Figure 8-20 (c). While soldering, small solder drops may appear behind the point of solder application, indicating the joint is full to that point and will take no more solder. Because these drops indicate that solder in the upper part of the joint is above the pasty temperature range and in the full liquid state, we want to cool the joint slightly to get back into the pasty temperature range. To do this, relocate the flame to the base of the fitting and apply more solder. Adding this solder will also cool the joint. Throughout this process you are using all three physical states of solder: solid, pasty, and liquid. Remember that you want to control the state of the solder: cooler solidified solder at the bottom of the fitting, pasty transitional state above the solid, and liquid solder at point of application to allow the solder to be drawn into the fitting. Beginners often apply too much heat and cannot get the solder to stay inside the fitting because it is too liquid.

For joints in the vertical position:

- After preheating, make a series of overlapping circular passes feeding in solder and starting wherever is convenient. Stop applying solder when you can see the joint is filled and it will take no more solder. Vertical joints are the easiest to start on.
- Since vertical joints are the easiest to make, it is often wise to plan the job so the maximum number of joints are vertical. Plumbers often preassemble sections of tubing and fittings on a bench or in a vise where it is more convenient and comfortable to work.

For fittings with horizontal or cup-inverted joint:

- Begin soldering the bottom-most joint first since the heat applied to it will preheat the joints above it. If you solder the upper joints first, you will melt the solder out of them when you begin soldering the lower joints as the convected heat rises. By soldering the lowest joint first, you not only preheat the joint(s) above it, but you have total control of the heat input to this joint. If you overheat this joint, the solder will be too thin (runny) to be held in place by capillary attraction, and it will run out. Just the right amount of heat will let the solder be drawn up into the joint. After the lowest joint is done, the side or top joint will then be ready to solder. Apply a little more heat to it, and run the solder around the ring where the tubing meets the fitting on the side, then the top joint.
- In all cases, trying to improve the integrity of a soldered joint by loading additional solder where the tubing enters the fitting is pointless. The strength of the joint is the solder bond between the inside of the fitting and the outside of the tubing inside the fitting. If the joint was clean, properly fluxed and heated when the solder was applied, it will hold.

Figure 8-21. A perfect male joint

186

Finishing the Job

- First, allow the assembly to air cool without disturbing it. Do not spray it with water or dip the fitting in water as cracking may result.
- Then, once the joint has reached room temperature, take a clean, cotton rag dipped in water or alcohol and wipe the flux residue from the joint. Wiping the joint while the solder is still molten can lead to joint cracking and is not recommended. Excess solder should be removed with the torch while the joint is still hot. Figure 8-21 shows what a perfect male joint should look like.
- Finally, test the completed joint for leaks, and flush the tubing lines to remove the remaining flux.

Soldering Tubing to a Valve

When soldering tubing to a shut-off valve, it is best to remove the vale stem before beginning work because heat form the torch can damage washers and other packing in the valve. Remove the stem by turning the packing nut counterclockwise with a wrench. Some ball valves contain Teflon seals that can withstand soldering temperatures for a short time, so there is no need to disassemble the vale.

Dealing with Water in the Tubing

Soldering must be done in tubing and fittings that are free of water.
Some methods to achieve a dry pipe when the line has been in service are:

- Open taps above and below the joint to help the tubing drain.
- Remove pipe straps on each side of the joint to permit bending the tubing down to let water run out.
- On vertical lines containing water, use a short length of 1/4-inch hose to siphon out the water so the water level inside the line is at least 8 inches below the joint.
- Using paper towels and a stick, dowel, or long screwdriver, remove all moisture and water drops within 8 inches (200 mm) of the soldering.
- Whenever you see steam rising from a fitting after sweating, there is a very good chance that water in the lines has spoiled the joint. Take it apart, fully drain the lines, and begin again.
- Valves on the lines to be soldered may not shut off completely, or it may be impossible to drain the line enough to prevent all water from entering the soldering area. To keep this water away, stuff fresh white bread (cut off crusts as they dissolve slowly) into the line(s) to form a plug and push it 8 inches (200 mm) back into the line away from the fitting. Then assemble the copper tubing to the fitting and complete the solder joint. Flush out the bread when the joint is complete.

187

What are some troubleshooting tips for soldered and brazed joints?

Although soldering and brazing operations are inherently simple, the omission or misapplication of any single part of the process may mean the difference between a good joint and a failure. Faulty joints usually result from one or more of the following factors:

- Improper joint preparation prior to soldering.
- Lack of proper fitting and tubing support during soldering or brazing may cause wiggling of the joint while the filler metal is molten. Non-concentric joint alignment may lead to gaps in filler metal.
- Improper heat control and heat distribution through the entire joining process.
- Improper application of solder or brazing filler metal to the joint.
- Inadequate filler metal applied to the joint.
- Sudden shock cooling and/or wiping the molten filler metal following soldering or brazing.
- Pretinning of joint prior to assembly and soldering.

Protecting Flammable Surfaces

Use a commercial woven glass heat shield to keep the torch flame off the building materials, Figure 8-22, or use a piece of heavy steel or aluminum to do the same thing. Old timers used pieces of asbestos and these worked even better. There are also commercial products in paste and spray form that help control heat from soldering and brazing operations. Examples of these products are *Block-it Heat Absorbing Paste* and *Cool Gel Heat Barrier Spray* from LA-CO®/Marcal.

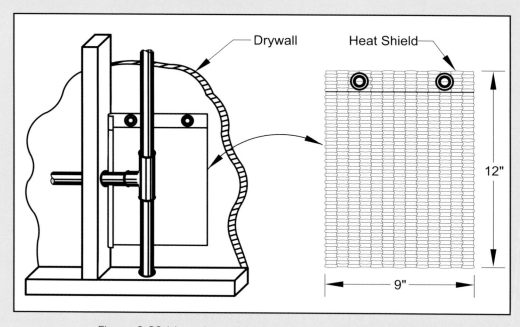

Figure 8-22 Use a heat shield when working in close quarters

As mentioned earlier in this chapter, brazing and soldering differ from welding in that the heat produced when brazing or soldering does not melt the base metals. Rather, the heated work draws a filler material into the joint through capillary attraction. Think of attaching a fitting to the end of a copper tube. In braze welding, the joints of two pieces of metal are open to the welder and filler material is deposited into the joints joining them together.

189

Figure 8-23 To begin braze welding, clean the metal thoroughly. For metal over 1/8 inch thick, grind a 46 to 60 degree bevel in both sections. This produces a wide groove that is larger than the grove used in most welding projects. Thin metals like those shown here do not need a groove for proper adhesion

Figure 8-24 Hold the torch about ¼ inch away from the work. The goal is to heat both sides of the joint evenly. Circle the torch and begin adding the filler material. Judge your speed by how well the filler material flows and adheres to the metal.

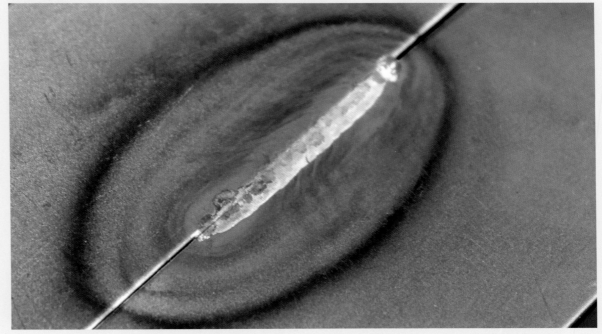

Figure 8-25 Here is a finished section of a brazed welded joint. The braze material should flow smoothly into the base metals on both sides of the joint. For added strength, you can make another pass over the first bead.

Welding Tasks, and Tips

This chapter contains instructions for performing some of the most common welding tasks. There are details on how to make rectangular frames, put legs on a table, and make a three-dimensional solid frame. There are also instructions for repairing cracked truck frames and welding on vehicles. In addition, there is a section on pipe and tubing repairs and tools. And there are expert tips on controlling distortion while welding.

WHAT'S THE BEST CORNER TREATMENT FOR A WELDED RECTANGULAR FRAME?

WHAT'S THE BEST WAY TO CHECK IF CORNERS ARE SQUARE?

HOW SHOULD A CRACKED C-CHANNEL TRUCK FRAME BE REPAIRED?

WHAT ARE SOME TIPS FOR WELDING PIPE AND TUBING?

WHAT ARE SOME WAYS TO DEAL WITH METAL DISTORTION?

What's the best corner treatment for a welded rectangular frame?

Mitering and notching are two common ways to make angle iron corners for a rectangular frame. See Figure 9–1 and 9–2. Both methods work, but a beginner might find notching easier because it is more dimensionally tolerant. Following welding and grinding either fit-up style will result in a good finish.

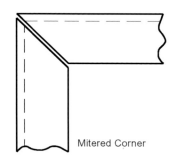 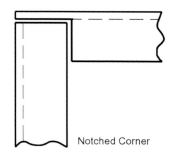

Figure 9–1 and 9–2
A mitered (left) and notched
(right) frame corners

Mitered Corner   Notched Corner

Single-Piece Frame

To make a rectangular frames from a single length of angle iron, see Figure 9-2. This approach works well when a notching machine is available and lends itself to production work. Getting the correct bend allowance gap is critical, because it provides the extra material needed to go *around* the outside of the corner when the bend is made in Figure 9–3(a). Begin by setting the bend allowance gap to slightly less than the thickness of the angle iron and go from there.

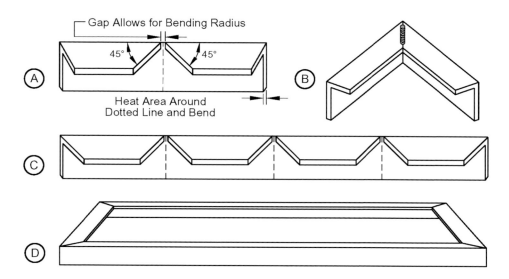

Figure 9-3 Notching and bending to make a single-piece frame:
(a) Corner detail before bending, (b) corner detail after bending, welding, and grinding,
(c) notched angle iron frame ready for bending, and (d) completed frame

What's the best way to check if corners are square?

Check for equal diagonals between opposite corners with steel measuring tape. On large frames use a carpenter's square, on smaller ones use a machinist's square. If the sides of the frame are to be plumb and level, a large level can be used. When welding a very large L-shape, where a square is too small and there are no diagonals to measure, use a 3-4-5 triangle methos: (1) measure off four units (feet, meters or some multiple of same) on one leg, and (2) measure off three units on the other leg, (3) then adjust the hypotenuse, the longest side of the triangle. This procedure makes a perfect right triangle. See Figure 9–4.

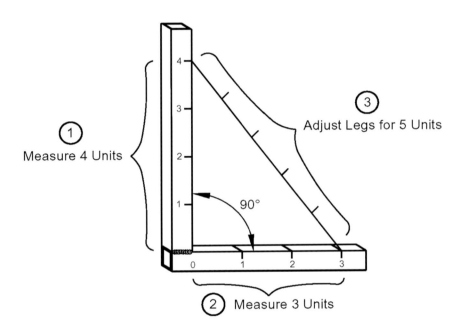

Figure 9-4 How to utilize a 3-4-5 triangle to set members at a right angle

Tips for Welding Square Frames

In *decreasing* order of effectiveness:

- Secure members in a rigid fixture and weld them in it.
- Clamp members to a steel table and then weld them.
- Use a fixture to hold parts for tacking, then weld the tacked parts *outside* the fixture. This fixture can be as simple as a sheet of plywood with wood blocks fixed to it to hold the work in place while the tack welds are made.
- Use Bessy®-type corner clamps. Hint: Begin by tack welding each of the corners together using the clamp each time; then check for square. Bend back into square if needed. If the tack welds are not too large, you'll be able to straighten the frame with moderate force— by hand and without hydraulic jacks. Begin final welds at *opposite* corners. **Warning:** Making a final weld one corner at a time in a corner clamp will bring poor results— the final two corner pieces are not likely to meet.
- Use magnetic corner tools—these are effective only for light sheet metal as they lack the strength to resist weld-induced distortion forces even with light angle iron, Figure 9-5.

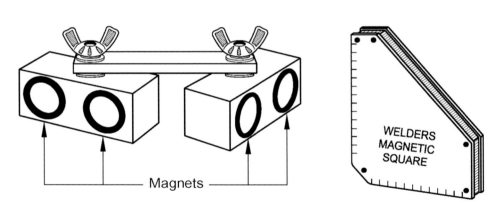

Figure 9-5 Magnetic corner tools

In the field or with large and heavy members, lay the members up square and level on a concrete floor (you may need to shim them to get them flat), tack them together; then weld them. Check for frame to be square and flat after making each tack, and bend members back to square and flat *before* making the next tack or weld. Large (and unobtainable) forces may be needed to bring the frame back into square with this method if your tacks are too large. **Warning:** Welding directly on concrete can cause it to explode violently. Shimming the work off the floor will eliminate this hazard.

To control distortion, weld opposite corners first, weld the *same* relative corner or side position in the exact same sequence on all four corner joints: First weld all outside faces, then all top corners, finally all bottom faces. Make each weld in the same relative direction: from the outside of the frame to its inside or vice versa. Also, give each weld a moment to cool before making the next.

Troubleshooting a Frame

If you have a welded a rectangular frame of angle iron (not rectangular tubing) that does not lie flat, follow the steps in Figure 9-6, which show how to bend the horizontal face of the frame to flatten it. Use an open end wrench or fabricate a tool of your own.

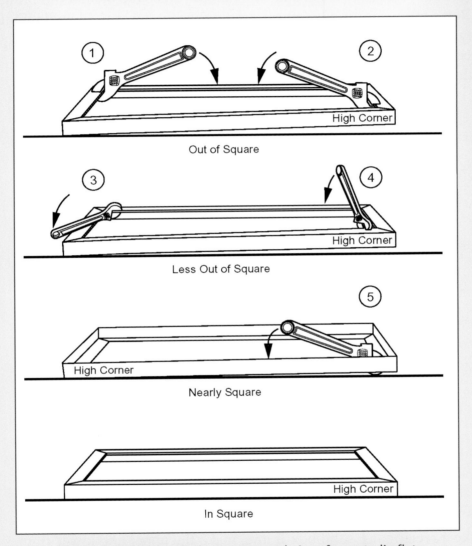

Figure 9-6 Method of adjusting an angle iron frame to lie flat

Mounting Table Legs

To weld supports to a rectangular frame—as in making a table—follow the steps in Figure 9-7.

Put the table frame upside down on a flat surface like the top of the welding table. Use two clamps to lightly secure a leg to both sides of the frame corner.

Using a carpenter's square, adjust the leg so that it is perpendicular to the frame. Using a length of steel or wood and two clamps, brace the member to bring it into square. Repeat this squaring/bracing/clamping for the other right angle, Figure 9-7 (b). Fully tighten two clamps holding the leg to the frame. Re-check for square in both directions, and adjust as needed; then weld the leg to the frame, Figure 9-7 (c). Repeat this for each leg.

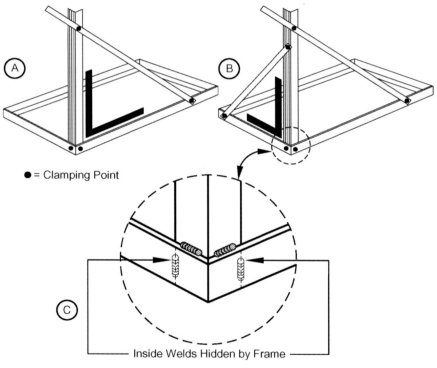

Figure 9-7 Welding table legs on square

Box Frames

To make a rectangular solid box frame, make the upper and lower frames as described previously. Use clamps to secure the four verticals to the lower frame, making them square to the lower frame, Figure 9-7. Place the completed upper frame over the legs. Make whatever compromises and adjustments are needed to the verticals to make them meet the upper frame. Some tweaking may be necessary. Note that all opposite diagonals, like dotted line X-Y, Figure 9-8 will be the same length in a *rectangular* box. Clamp the legs to the upper frame. Tack all of the joints, keeping the components square. Then weld all joints. This method will work equally well with angle iron or rectangular tubing. See Figure 9-8.

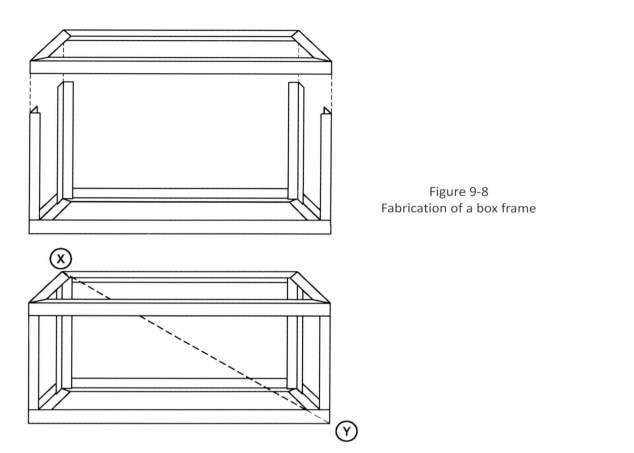

Figure 9-8
Fabrication of a box frame

197

How should a cracked C-channel truck frame be repaired?

Many of the C-channel frames of modern tractors and heavy trucks carry a label that warns against cutting or welding on them. To save weight, manufacturers used thinner, lighter steel U-channel members with a special heat treatment to provide extra strength. Welding and torch cutting on these members destroys the strength of the factory heat treatment. Do not weld, flame cut, or drill on these members if they have not failed. If you have to mount something on the frame, use the extra and unused existing holes put in at the factory. However, if C-channels must be repaired, minimize welding on them.

Use the following steps:

Clean the repair area. First, steam clean and scrub the entire area surrounding the weld. (This cleaning is particularly important for waste hauling vehicles.) Then use an oxy-fuel torch to dry this area and remove remaining mill scale. Finally, wire brush the area down to shiny metal. Compare your failure with those shown in Figure 9-9 to determine which example your frame failure matches best, and then follow the repair steps for that example

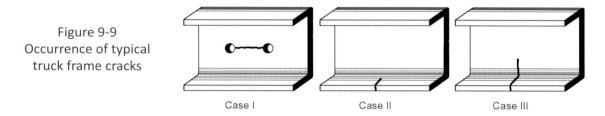

Figure 9-9
Occurrence of typical
truck frame cracks

Case I   Case II   Case III

Case I—Horizontal crack along the web between factory-drilled holes.

Grind a V-groove to within 1/16 to 1/8 inch of the thickness of the web steel along the path of the crack and extending 2 inches beyond the initial crack on each end. Put this V-groove on the inside of the C-channel, Figure 9-10 (b). Use a copper backing plate clamped to the back of the groove to protect the back of the weld from atmospheric contamination. Using SMAW low-hydrogen or iron powder electrodes, fill the V-groove with weld metal and grind it flush. Use enough current for a full penetration weld. See Figure 9-10 (c). Grind the weld flush (on the outside of the channel too if necessary to make it flat) making sure to leave the ground surface as smooth as possible. Any irregularities or scratches are stress raisers. Cut and fit a ½-inch-thick carbon steel reinforcement plate on the inside the web extending at least 6 inches beyond the ends of the weld repair. Grind this plate on the lower and upper edges so it fits *tightly* against the web and

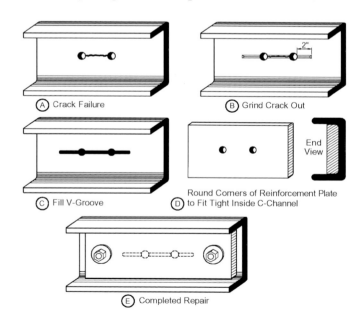

Figure 9-10 Case I: Horizontal crack in truck C-channel between factory-drilled holes in web. This is a common case and cracks as long as 10 inches (250 mm) can occur

edges of the flanges for its entire distance, Figure 9-10 (d). Using existing factory-drilled C-channel holes, if possible, secure the reinforcement plate to the web with bolts matching the diameter of these holes. Holes are usually sized 1/2, 9/16, or 5/8 inch in diameter. Use a washer under each nut. If no holes are available, drill your own. Stop. The repair is complete. No additional welding is needed. See Figure 9-10 (e).

Case II—Crack on bottom flange perpendicular to web only.

Here are the repair steps:

Grind a V-groove half way through the thickness of the flange along the path of the crack. See Figure 9-11 (b). Using SMAW with a low-hydrogen or iron powder electrode, fill the V-groove with weld metal and grind it flush. Use enough current for full penetration of the flange metal. See Figure 9-11 (c). Grind the weld flush (on the inside of the flange too if necessary to make it flat) making sure to leave the ground surface as smooth as possible. Any remaining surface imperfections are stress raisers. Cut a 1/2- by 1 1/2-inch reinforcement bar from mild carbon steel 12 to 15 inches (300 to 380 mm) long. Center it on the crack. Weld this bar on the middle of the flange using a SMAW low-hydrogen or iron powder electrode. The 1 1/2-inch dimension of the bar is vertical. See Figure 9-11 (d). Using a grinder, gouge a bevel groove through the unwelded side of the reinforcement bar to sound weld metal on the other side, Figure 9-11 (d). Place a fillet weld in the gouged groove using SMAW with a low-hydrogen or iron powder electrode to secure the other side of the reinforcement bar. See Figure 9-11 (e). Do *not* make welds perpendicular to the length of the channel at the ends of the reinforcement bar.

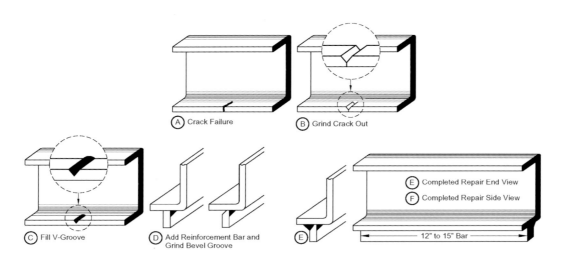

Figure 9-11 Case II: Crack on bottom flange perpendicular to web only

Case III—Crack on bottom flange perpendicular to web and extending up into web.

This is what happens when Case II is left uncorrected. See Figure 9-12. The repair steps are:

- Grind out crack with a V-groove halfway through the thickness of the flange along the path of the crack.
- Drill a 3/8- to 1/2-inch (10 to 13 mm) diameter hole at the end of the crack in the web to relieve stress.
- Use reinforcement methods of Cases I and II to add both a bottom reinforcement bar and a reinforcement plate inside the channel web.

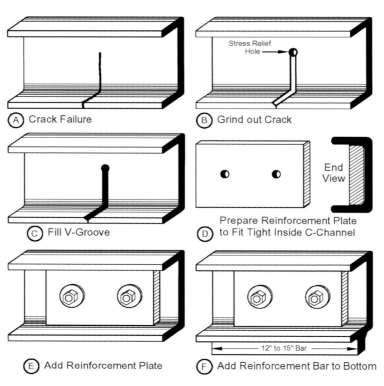

Figure 19-12 Case III: Crack begins on bottom flange and extends into web of channel

Traditional C-Channel Repair

Another and more traditional C-channel repair approach is to weld reinforcement plates to the web instead of bolting them or to weld a reinforcement bar along the bottom flange, but this is less commonly done today. If you choose welding instead of bolting for C-channel web repair, be sure to place welds *parallel* with the channel. Do not place any welds perpendicular to the channel. Welds perpendicular to the channel concentrate stresses on just one section of the weld. This is because the end welds on the patch plate prevent beam stress from being distributed evenly along the weld length. They become a new stress raiser and will produce near-term failure. See Figure 9-13.

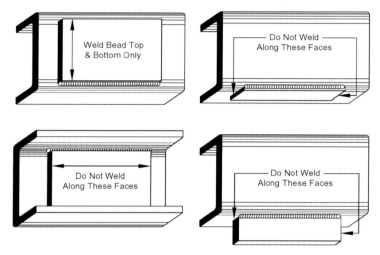

Figure 9-13 Welding reinforcement plates to C-channels: Welds parallel to C-channels are acceptable; perpendicular welds are not

Caution: Remember that these failures occurred in the first place because the member was stressed beyond its design capacity by excessive loads and fatigue. Dump truck body action stresses, hydraulic cylinder loads, road vibration, and truck overloading all contribute to failure. *Failure is likely to* happen again, usually in the next weakest location because the member is subject to the same load conditions that caused the initial failure.

Welding High-Strength Steel

Oxyacetylene welding on high-strength sheet metal on newer model cars leads to cracking. Instead use MIG welding.

Begin the weld bead on the outside edge of the crack and work toward the *inside,* which will keep the inherent weakness of the bead-ending crater away from the metal's edge where it would act as a stress raiser and lead to a new failure. On galvanized body parts ER70S-3 electrodes should be used as they contain less silicon than ER70S-6 electrodes which contributes to cracking when mixed with zinc.

What are some tips for welding pipe and tubing?

It is important to understand the difference between pipe and tubing. Pipe usually has a much thicker wall than tubing. These thicker walls permit pipe to accept threads; tubing because it has thinner walls, cannot be threaded. Another difference is that pipe from 1/4 to 12 inches (6 to 300 mm) diameter is specified by its *inside* diameter; pipe 14 inches diameter and larger is specified by its outside diameter. Tubing is always specified by its *outside* diameter and wall thickness.

Copper tubing is available in *drawn* or *annealed* condition. Drawn copper tubing cannot be bent (unless it has been annealed) without having its sidewalls collapse. It comes only in straight lengths of 12, 18, and 20 feet and is ideal for straight runs where appearance is important. Annealed tubing comes in both coils and straight lengths and can be easily bent without tools in smaller diameter sizes without wall collapse. Because it can be formed to bend or fit around obstacles along its path, many fittings (and copper soldering joints) can be eliminated by using a *single* piece of annealed tubing. There are three common wall thicknesses: Type K (heaviest), Type L (standard), and Type M (lightest). All three are used in domestic water service and distribution. Budget, preference, and local codes govern this choice.

Pipe Cutting Methods

For most jobs, you will need to cut pipe and tubing to length, see Table 9-1. Once cut, prepare the pipe or tubing by cleaning, grinding, brushing, or beveling the ends.

Table 9-1 Cutting methods for carbon steel pipe

Type Pipe	Method of Cutting
Steel pipe 3" (80 mm) diameter or smaller	Manual wheeled pipe cutter, or Wheeled pipe cutter operating in powered threading machine, or Portable bandsaw, or Oxyfuel cutting
Steel pipe over 3" (80 mm) diameter	Oxyfuel cutting, or Abrasive cut-off saw

How should steel pipe be cut with an oxyfuel torch?

In cutting small diameter pipe, the torch remains in the vertical position at all times. First, a cut from the 12 o'clock to the 9 o'clock position is made, and then another cut from the 12 o'clock to the 3 o'clock position is made. The pipe is rotated a half-turn and the process repeated. See Figure 3–9.

Cutting Large-Diameter Dipe

To mark the cut line on large pipe, use a *wrap-around,* which is a length of thin, flexible material—vinyl, fiber, or cardboard—that is wrapped around the pipe 1 1/2 times as shown in Figure 9–14. Adjust the wrap-around so that the second, overlapping layer lies squarely over the first layer, then hold the wrap-around tight with one hand and mark the cut line with the other. A typical wrap-around is 1/16 inch thick by 2 1/4 inches wide by 48 inches long (1.6 mm thick, 6 cm wide, by 120 cm long).

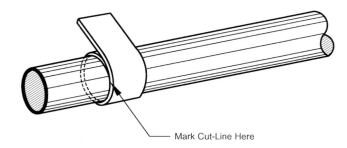

Mark Cut-Line Here

Figure 9–14 Utilizing a wrap-around to mark the cut will help in making a square cut line

Joining Lengths of Pipe

If you need to join two lengths of pipe and want to align them accurately, but you do not have a commercial pipe welding fixture, tack two lengths of angle iron to form a double V-base as in Figure 9–15. Many applications will require a longer welding fixture than the one shown. Align the pipes before tack welding, make the root pass, followed by subsequent passes.

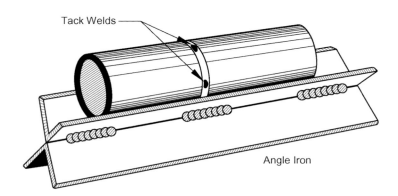

Tack Welds

Angle Iron

Figure 9–15 Using angle iron to align pipe

Working with Pipe

Pipe rollers permit the welder to always be welding in the flat position. The welder welds while a helper rotates the pipe to maintain the weld area on top of the pipe so welding is done in the flat position, or in field welding the welder may make weld in a fixed or stationary position. Using four rollers together keeps the pipes in alignment while they are being welded. On very large pipe or castings, a motor driven positioning devise may take the place of rollers. See Figure 9–16.

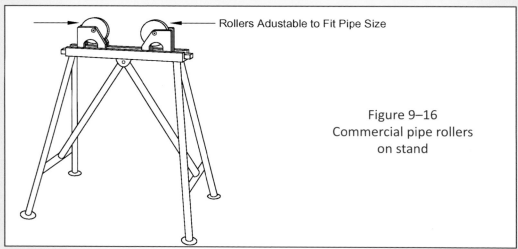

Rollers Adustable to Fit Pipe Size

Figure 9–16
Commercial pipe rollers
on stand

A Curv-O-Mark® Contour tool shown in Figure 9–17 locates any angle on the outside of the pipe. Just set the adjustable scale to the angle desired, place the tool on the pipe to the "level" position of the bubble, and strike the built-in center punch to mark the point. Figure 9–17 shows the contour tool locating the exact top, or zero degree position, the 90 degree position and the 60 degree position.

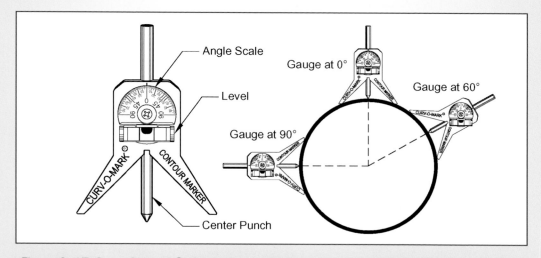

Angle Scale

Level

Center Punch

Gauge at 0°

Gauge at 60°

Gauge at 90°

Figure 9–17 Curv-O-Mark® Contour tool for locating angular positions around a pipe

Cracks in Structural Tubing

To repair cracked *structural* tubing (carries no fluid), follow these steps.

• Drill 1/4- inch (6.5 mm) stress relief holes at the ends of the crack.
• Cut or grind away the cracked area between the 1/4-inch holes.
• Cut a patch out of a similar-sized piece of tubing, and reshape this patch to fit over the cracked tubing area.
• Weld on the patch as shown in Figure 9–18: Use no continuous welds, no end welds and no welds closer to the end of the patch piece than 1/4 inch (6.5 mm).

205

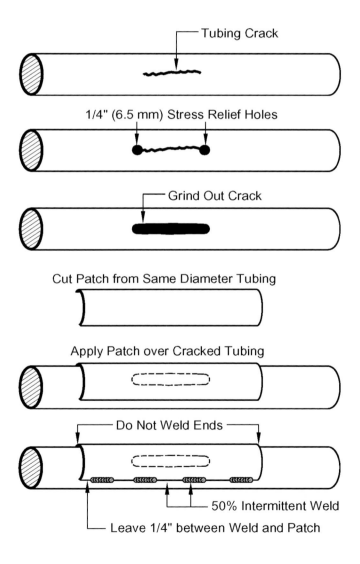

Figure 9–18 Repairing cracked structural tubing

Joining Pipe

Here are four ways to join one pipe inserted into a larger pipe by welding for a *structural* application (no fluid carried).

See Figure 9–19.

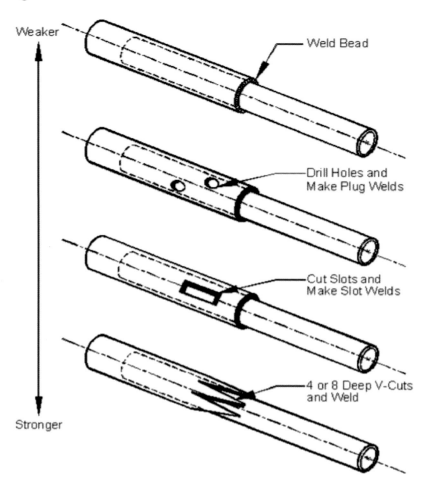

Figure 9–19 Ways to join pipe lengths by welding

Equal Diameter Pipe. To join two equal-diameter pipes in a *structural* application that will be subjected to torsion, tension, or shear. Use the method in Figure 9–20 to ensure that the pipes are in alignment; then make a full penetration weld. Such a weld is as strong as the pipe itself. Note that the inner pipe is for proper alignment, not strength.

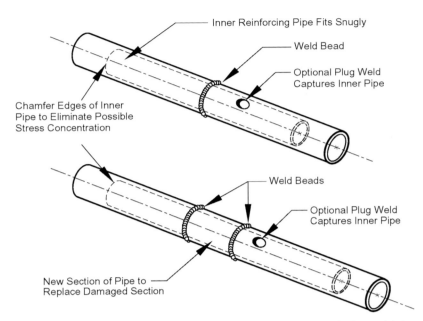

Figure 9–20 Two different methods of splicing structural pipe which assure concentric alignment

Tubing Saddle

A saddle (or *fishmouth*) is the shaping of the end of one piece of tubing so it meets and fits tightly against another. We do this to make strong welds. The gap between the two tubing pieces should not exceed the diameter of the MIG or TIG welding wire used to make the weld. See Figure 9–21.

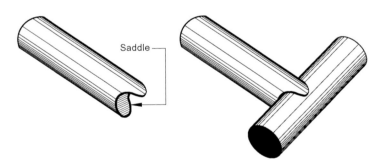

Figure 9–21 Tubing saddle

You can use hack saws and files to make a saddle. Begin by forming the saddle on a bench or pedestal grinder, making the final adjustments with a file. A hand-operated nibbling tool is a faster way to make a saddle, although some operator skill is needed. Using a saddling hole-saw tool works well.

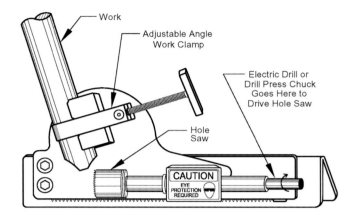

Figure 9–22 Commercial saddle cutting tool

Protecting Thin Tubing

To protect thin-walled tubing while holding it in a vise, first insert a tightly-fitting wooden dowel inside the tubing end to prevent it from collapsing in the vise. Then place soft jaws over the steel jaws of the vise to prevent marring the tubing. The soft jaws are usually made of aluminum, copper, lead, or plastic. See Figures 9–23 and 9–24.

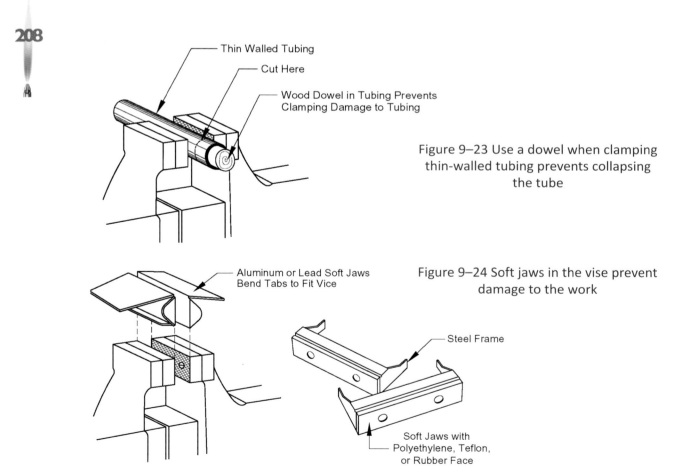

Figure 9–23 Use a dowel when clamping thin-walled tubing prevents collapsing the tube

Figure 9–24 Soft jaws in the vise prevent damage to the work

What are some ways to deal with metal distortion?

Distortion is the permanent change in shape and dimension of metal caused by expansion and contraction. Certain welding and cutting processes are sources of distortion. It is a result of *residual stress* left by uneven heating that causes permanent shape changes. This stress not only can ruin the shape of a part, it can weaken it too. Even through the part bears no external load, the residual stress acts as an initial load on top of what is externally imposed and reduces the total load the part can withstand.

Because welding processes expose the workpiece to high temperatures, weld-induced distortion is always present. We will investigate how distortion affects workpieces and present several methods to reduce its effects. These methods are often simple, but without them many workpieces would be ruined.

Examples of Distortion

Before it is heated the sheet metal is square and flat. As one edge of the plate heats, it expands and softens while the cool edge does not. When the sheet cools back to room temperature, most wrinkles disappear, but the sheet is permanently shorter along the once-heated edge. Uneven heating of the plate and the restraint offered by its cooler side cause a dimensional change called *upsetting*. Upsetting will be even more severe if a water spray cools the sheet rapidly and prevents the unheated edge from heating and expanding as the heat flows across the plate. Note that what happens during the cooling period is seldom what happens during the heating period. See Figure 9–25.

209

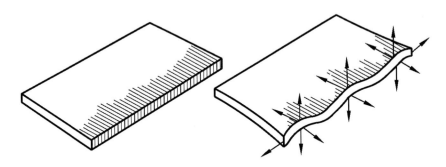

Figure 9–25 Sheet metal before heating (left) and while hot (right)

Distortion Produced by Oxy-fuel Cutting. Because of differential heating and the restraint offered by the uncut portion of the plate, the *hinge effect* occurs, see Figure 9–26. If two parallel cuts were made simultaneously, the metal between the cuts would show little distortion, as heating and expansion would be balanced. Dual torches are available to make such cuts.

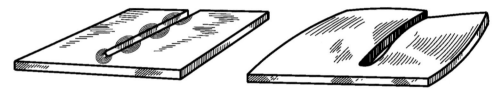

Figure 9-26 Hinge effect in partially cut plate

Weld Bead Distortion. See Figure 9–27. As the bead is applied, one side of the bar is heated and expands. When the bar and the filler metal on top of it cool, they contract much more than the cooler metal on the bar's opposite side. Applying the weld bead makes the bar bend toward the side of the weld bead.

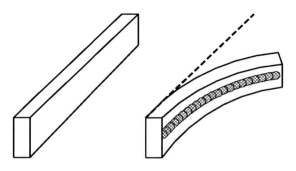

Figure 9–27 Affect of applying a weld bead on one side of a steel bar. Straight bar, no bead applied (left). Bar bends toward bead side with weld bead applied when cool (right)

V-Groove Butt Weld Distortion. There are residual stresses both along the weld axis (longitudinally) and across the weld (transversely), Figure 9–28 (left). When cooled, the weldment has permanent deformation away from the side where heat was applied along the weld line. Because this is a V-groove butt joint that has more filler metal at its top than at its root, there is more shrinkage along the top of the joint in both directions. This makes the plate dish or bend upwards, Figure 9-28 (right).

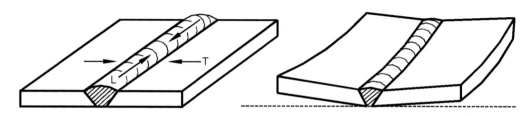

Figure 9–28 Butt weld longitudinal, L, and transverse, T, residual stresses (left) and resulting deformation (right)

T-Joint Distortion. See Figure 9–29 (left) showing both longitudinal and transverse stresses in the weld bead. There is a second weld bead on the back side of the T-joint. Because the longitudinal stresses on each side of the joint balance each other, the vertical member of the T-joint remains straight. See Figure 9–29 (right) for the distortion these residual stresses cause.

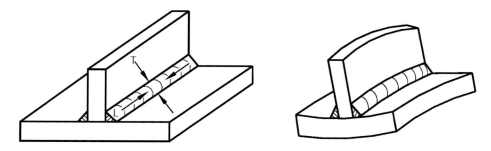

Figure 9–29 Longitudinal and transverse stresses in a T-joint (left) and the distortion they cause (right)

Controlling Distortion

There are several steps that can reduce the effects of distortion, but we can never completely eliminate distortion.

211

Butt Joints and V-Groove Joints. First, preset the parts. Then tack-weld the parts slightly out of position and let residual forces bring them into proper position. See Figure 9–30 showing how a T-joint and V-groove joint are handled.

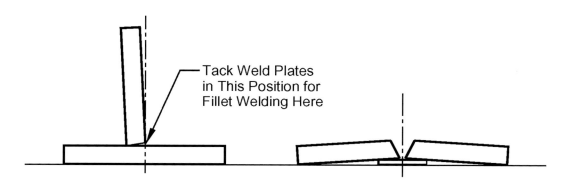

Tack Weld Plates in This Position for Fillet Welding Here

Figure 9–30 Presetting and tack welding work out of position let weld shrinkage bring parts back into alignment

By clamping, the use of restraints and wedges hold the weld joint in proper position until the weld metal cools. This approach may not produce perfect results, but it will help reduce distortion. See Figure 9–31.

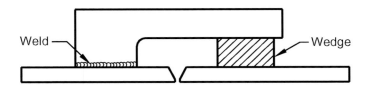

Figure 9–31 Use tack-welded restraints and wedges to hold joint in position while welding

Limiting Heat Flow. Use chill bars. Chill bars consist of steel or copper bars clamped beside and parallel with the weld bead. They draw heat away from the weld and reduce its flow to the rest of the part. They also limit distortion to upsetting metal close to the weld line and eliminate ripples completely by exerting a clamping force which prevents ripples from forming when the work is hot, see Figure 9–32. A groove in the lower chill bar permits the weld itself to remain hot and not have its heat drained away by the chill bar. This groove could be flooded with shielding gas for GTAW.

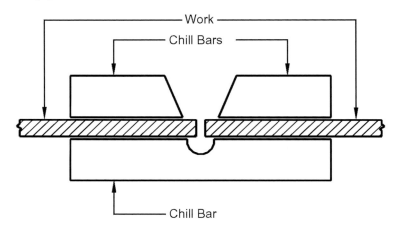

Figure 9–32 Chill bars reduce distortion by confining heat to the weld area and by preventing work from forming ripples when hot

Prestressing. Use clamps to bend the joint members in opposite direction to the weld forces, letting weld-shrinkage forces bring the parts back into position. This method works well when a jig or fixture can be used and test runs are made to determine the amount of prestress needed. See Figure 9–33 on following page.

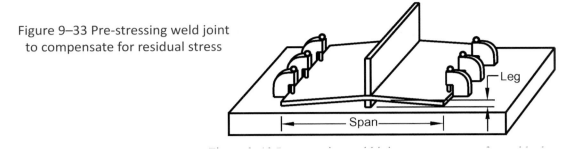

Figure 9–33 Pre-stressing weld joint to compensate for residual stress

Balance Distortion Forces. Use equal distortion forces to balance each other by using two (or more) weld beads. This could be done by putting a fillet weld on both sides of a T-joint or using a double V-groove butt joint. See Figure 9–34.

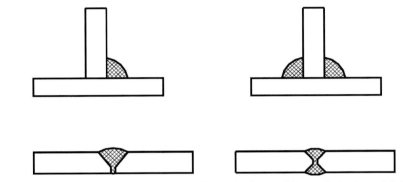

213

Figure 9–34 Initial joint designs (left) and balanced forced designs (right)

Use chain intermittent or staggered intermittent weld beads. Intermittent beads not only balance one another, but also by reducing the total amount of weld bead, reduce total residual force, see Figure 9–35. Even a single intermittent weld bead will have less distortion than a single continuous weld bead and often the strength of a continuous bead is not needed.

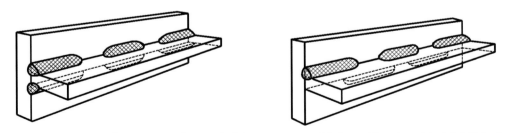

Figure 9–35 Using chain intermittent (left) or staggered intermittent welds (right) to balance forces and reduce total weld bead metal

Redesign Joints. Use a V-groove and a fillet weld in place of a fillet weld alone to balance residual stress. See Figure 9–36.

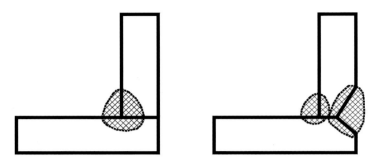

Figure 9–36 Redesigned joint can balance residual stress and reduce distortion

Long Continuous Beads. To reduce distortion, use back-step welding. Apply short increments of beads in the direction opposite of the end point of the weld. When applying multiple passes, start and stop the beads of each layer at different points. See Figure 9–37.

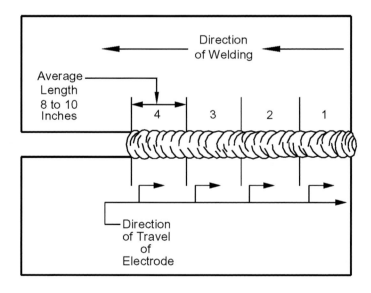

Figure 9–37 Back-step welding sequence

The use of wedges ahead of the weld will control joint spacing. See Figure 9–38.

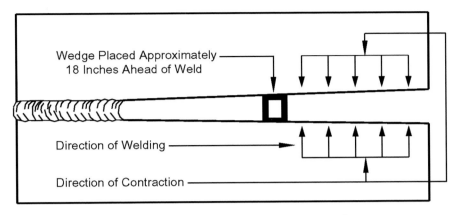

Figure 9–38 Control joint spacing with wedges

Other Techniques to Reduce Distortion

Preheating base metal—raising the temperature of the entire part before welding reduces temperature differences, residual stress, and distortion.

Peening—hammering the weld metal, usually with an air hammer, slightly reshapes the metal and redistributes concentrated forces. In a multipass weld, this is done between each pass. This method can be helpful, but depends on the skill and judgment of the welder; peening consistency is difficult to control.

Stress relieving heat treatment—using an oven or electric heating coil, the entire part or the weldment area is heated high enough to remove weld-induced stress. This is commonly done in structural steel work.

Brazing or soldering instead of welding—because brazing and soldering expose the workpiece to much lower temperatures than welding, these two processes can be used when the strength of welding is not required.

GLOSSARY

A

acetone: A colorless, flammable, volatile liquid used as a paint remover and as a solvent for oils and other organic compounds. Used in acetylene cylinders to saturate the monolithic filler material to stabilize the acetylene.
acetona: *Líquido incoloro, volátil, que se usa para remover pintura, y como disolvente de aceites y de otras substancias orgánicas. Para estabilizar el acetileno, se usa en las bombonas para que sature el material de aportación.*

acetylene feather: The intense white, feathery-edged portion adjacent to the cone of a carburizing oxyacetylene flame.
pluma de acetileno: *En una llama carburante de un soplete oxiacetilénico, la zona muy blanca, en forma de pluma, adyacente al cono de la llama.*

actual throat: The shortest distance between the weld root and the face of a fillet weld.
garganta actual: *En un cordón de soldadura, la distancia más corta entre la raíz de la soldadura y la superficie del cordón.*

adhesion: A state of being stuck together. The joining together of parts that are normally separate.
adhesión: *La condición de estar pegado. La unión de partes normalmente separadas.*

aluminum: One of the chemical elements, a silvery, lightweight, easily worked metal that resists corrosion.
aluminio: *Elemento químico, de color plateado, ligero de peso, fácil de trabajar mecanicamente, y muy resistente a la corrosión.*

ampere: A unit of electrical current measuring the rate of flow of electrons through a circuit. One ampere is equivalent to the current produced by one volt applied across a resistance of one ohm.
amperio: *La unidad de corriente eléctrica que mide el flujo de electrones en un circuito eléctrico. Un amperio corresponde a la corriente contínua producida por una tensión de un voltio, a través de una resistencia de un ohmio.*

annealing: A process of heating then cooling metal to acquire desired qualities such as ductility.
templaje: *El proceso de calentar y luego enfriar un metal para que adquiera ciertas propiedades, como la ductilidad, por ejemplo.*

anode: The positive terminal of an electrical source.
ánodo: *El polo positivo de una fuente de electricidad.*

arc cutting: A group of thermal cutting processes that severs or removes metal by melting with the heat of an arc between an electrode and workpiece.
corte con arco: *Un grupo de procesos térmicos para cortar metales. El metal se separa o se pierde fundiéndolo con el calor de un arco entre un electrodo y la pieza que se está cortando.*

alloy: A substance with metallic properties, composed of two or more chemical elements of which at least one is a metal.

aleación: Una substancia con propiedades metálicas, que está compuesta de dos o más elementos químicos, de los cuales al menos uno de ellos es un metal.

alloying element: Elements added in a large enough percentage to change the characteristics of the metal. Such elements may be chromium, manganese, nickel, tungsten, or vanadium; these elements are added to produce specific physical properties such as hardness, toughness, ductility, strength, resistance to corrosion, or resistance to wear.

elemento de aleación: Elemento que se añade a un metal en proporciones suficiente-mente altas para cambiar las características del metal. Algunos de estos elementos son cromo, manganeso, tung-steno o vanadio, etc. Estos elementos producen ciertas propiedades físicas como la dureza, la tenacidad, la ductilidad, la resistencia mecánica, y la resistencia a la corrosión y al desgaste.

alloy steel: A plain carbon steel to which another element, other than iron and carbon, has been added in a percentage large enough to alter its characteristics.

acero de aleación: Un simple acero al carbono al que se le añade un elemento – que no sea hierro o carbono – en cantidad suficiente para alterar sus propiedades.

alternating current (AC): An electric current that reverses its direction periodically.

corriente alterna (C.A.): Una corriente eléctrica que cambia de dirección periodicamente.

arc blow: The deflection of an arc from its normal path because of magnetic forces.

soplo del arco: La deviación de su trayectoria sufrida por un arco eléctrico debida a fuerzas mágnet-icas.

arc force: The axial force developed by an arc plasma.

fuerza del arco: La fuerza axial producida por el plasma de un arco.

arc gap: A nonstandard term used for the arc length.

abertura del arco: Una expresión, no aprobada, la cual se usa para designar longitud de arco.

arc gouging: Thermal gouging that uses an arc cutting process variation to form a bevel or groove.

estriar con arco: Escarbar un metal termicamente para formar un chaflán o una ranura, usando una variación del proceso de cortar metales con un arco.

arc plasma: A gas that has been heated by an arc to at least a partially ionized condition, enabling it to conduct electric current.

plasma de arco: Un volumen de gas que ha sido calentado por un arco a una temperatura tal que el gas está suficientemente ionizado para conducir una corriente eléctrica.

arc time: The time during which an arc is maintained in making an arc weld.

duración del arco: El tiempo durante el cual un arco es mantenido para hacer una soldadura por arco.

arc welding: Arc welding is a group of welding processes in which fusion is produced by heating with an electric arc or arcs with or without the application of pressure and with or without the use of filler metal.

soldadura con arco: Grupo de procedimientos para soldar, en los cuales fusión de los met-ales ocurre por medio de arco(s) eléctrico(s) con o sin la aplicación de presión y con o sin el uso de material de aportación.

arc voltage: The voltage across the arc.

tensión de arco: Diferencia de potencial a través del arco.

ASTM: The American Society for Testing and Materials.

ASTM: Organización en los EE.UU. llamada Sociedad Americana para Ensayos y Materiales.

autogenous weld: A fusion weld made without using a filler material.
soldadura autógena: Soldadura a fusión hecha sin usar material de aportación.

AWS: The American Welding Society.
AWS: Organización en los EE.UU. llamada Sociedad Americana para Soldadura.

axis of a weld: A line through the length of a weld, perpendicular to and at the geometric center of its cross-section.
eje de una soldadura: Una línea a lo largo de una soldadura, perpendicular a su sección, y que pasa por su centro geométrico.

B

back bead: A weld bead resulting from a back weld pass. Back beads are made after the primary weld is completed.
retroreborde: Un cordón de soldadura , el resultado de un pase invertido. Estos cordones son siempre aplicados después de haberse hecho la soldadura principal.

back fire: The momentary recession of the flame into the welding tip, or cutting tip followed by immediate reappearance or complete extinction of the flame, accompanied by a loud popping report.
retroquema: La desaparición momentánea de la llama del soplete dentro de la boquilla de soldar o de cortar. Esta desaparición, a veces, causa la extinción de la llama, y otras veces la llama reaparece inmediatamente, pero en ambos casos, está acompañada de una explosión muy sonora.

backgouging: The removal of weld metal and base metal from the weld root side of a welded joint to facilitate complete fusion and complete joint penetration upon subsequent welding from that side.
Escarbando por detrás: La remoción de porciones de metal de base y metal de soldadura del lado de la raíz, para promover una fusión completa y una penetración completa de la juntura al hacer un pase subsiguiente, de ese lado.

219

backhand welding: A welding technique in which the welding torch or gun is directed opposite to the progress of welding.
soldando con el dorso: Una técnica de soldar en la que se hace que el soplete o la pistola esté orientado en dirección opuesta a aquella en la que progresa la soldadura.

backing: Material or device placed against the back side of a joint to support and retain molten weld-metal. The material may be partially fused or remain unfused during welding and may be either metal or nonmetal (metal strip, asbestos, carbon, copper, inert gas, ceramics).
soporte: Un aparato o dispositivo que se monta en la parte trasera de la juntura para soportar y retener el metal de aportación fundido. El aparato puede a veces fundirse parcialmente, o puede permanecer separado durante la soldadura; y puede ser hecho de metal o de otra substancia (fibras metálicas, asbesto, carbón, cobre, un gas inerte, o cerámica).

backing bead: A weld bead resulting from a backing pass. Backing beads are completed before welding the primary weld.
cordón de respaldo: El cordón que resulta de un pase de soldadura de respaldo. Estos cordones se completan antes de empezar la soldadura principal.

backing pass: A weld pass made to provide a backing for the primary weld.
pase de respaldo: El pase de soldadura hecho en una soldadura de respaldo.

backing strip: Non-standard term used to describe a backing on the root side of the weld in the form of a strip.
tira de soporte: Un término, no aprobado, que se usa para describir un soporte, en el lado de la raíz, que tiene la forma de una tira.

back-step sequence: A longitudinal sequence in which weld passes are made in the direction opposite weld progression, usually used to control distortion.
secuencia "pase atrás": *Una secuencia longitudinal en la cual algunos pases se hacen en la dirección opuesta a la dirección en la que la soldadura avanza. Esta técnica se usa principalmente para controlar la deformación de la pieza.*

back weld: A weld made at the back of a single groove weld.
soldadura trasera: *Una soldadura hecha en la parte trasera de una soldadura con una ranura nada más.*

base material: The material that is welded, brazed, soldered, or cut.
material de base: *El material que va a ser soldado con arco,, soldado fuerte o débil, o cortado.*

base metal: The metal or alloy that is welded, brazed, soldered, or cut.
metal de base: *El metal o la aleación que va a ser soldado con arco, soldado fuerte o débil, o cortado.*

bead weld: A term used for surfacing welds.
soldadura a cordón: *Otro nombre para las soldaduras de alisamiento.*

bevel: An edge preparation, the angular edge shape.
chaflán: *La preparación, de forma angular, del borde de una pieza.*

bevel angle: The angle between the bevel of a joint member and a plane perpendicular to the surface of the member.
ángulo del chaflán: *El ángulo formado por el plano que contiene el chaflán de una de las piezas de una juntura y un plano perpendicular a la superficie de la pieza.*

220

boxing: The continuation of a fillet weld around a corner of a member as an extension of the principle weld.
encajonamiento: *La continuación de un cordón de soldadura en torno de una esquina de una pieza, como extensión de la soldadura principal.*

brazing: A group of welding processes that produces coalescence of materials by heating them to the brazing temperature in the presence of a filler metal having a liquidus above 850°F (450°C) and below the solidus of the base metal. The filler metal is distributed between the closely fitted faying surfaces of the joint by capillary action.
soldadura: *Un grupo de procesos de union que producen la coalescencia de materiales, aumentando sus temperaturas , en la presencia de un metal de aportación cuyo liquidus es más de 840°F (450°C) y menos que el solidus del metal de base. El metal de aportación se difunde entre las superficies estrechamente yuxtapuestas por medio de la acción de la capilaridad.*

brazing filler metal: The metal or alloy used as a filler metal in brazing, which has liquidus above 850°F (450°C) and below the solidus of the base metal.
metal de aportación para soldaduras fuertes: *El metal o la aleación que se usa como metal de aportación en soldaduras fuertes, el cual tiene un liquidus por encima de 840°F (450°C) pero por debajo del solidus del metal a soldar.*

buckling: Bending or warping caused by the heat of welding.
comba: *Flexión o arqueamiento causado por el calor producido al soldar.*

buttering: A surfacing variation that deposits surfacing metal on one or more surfaces to provide metallurgically compatible weld metal for the subsequent completion of the weld.
emplastar: *Una variación de alisamiento que deposita el metal de alisamiento en una o más superficies para crear compatibilidad metalúrgica en el metal de soldadura en las subsiguientes operaciones hasta completar la soldadura.*

butt joint: A joint between two members aligned approximately in the same plane.

junta a tope: *Una juntura de dos miembros alineados aproximadamente en un mismo plano.*

C

capillary action: The force by which liquid in contact with a solid is distributed between closely fitted surfaces of the joint to be brazed or soldered.

acción capilar: *La fuerza por la que un líquido en contacto con un sólido se distribuye entre las superficies estrechamente yuxtapuestas de una juntura que va a ser soldada.*

carbon: A nonmetallic chemical element that occurs in many inorganic and all organic compounds. Carbon is found in diamond and graphite, and is a constituent of coal, petroleum, asphalt, limestone, and other carbonates. In combination, it occurs as carbon dioxide and as a constituent of all living things. Adjustment of the amount of carbon in iron produces steel.

carbono: *Un elemento químico no metálico que se encuentra en muchos compuestos inorgánicos y en todos los compuestos orgánicos. En su estado natural se encuentra como diamante y como grafito, y es un componente del carbón, petróleo, asfalto, piedra caliza, y otros carbonatos. En combinación, se halla en el dióxido de carbono, y como constituyente en todas las substancias o cosas animadas. La añadidura de carbono al hierro produce el acero.*

carbon steel: A steel containing various percentages of carbon. Low-carbon steel contains a maximum of 0.15% carbon; mild steel contains 0.15% to 0.35% carbon; medium-carbon steel contains 0.35% to 0.60% carbon; high-carbon steel contains from 0.60% to 1.0% carbon.

acero al carbono: *Acero que contiene varios porcentajes de carbono. Acero con poco car-bono tiene un máximo de 0,15% de carbono; acero dulce tiene de 0,15% a 0,35% de car-bono; acero a medio carbono tiene de 0,35% a 0,60% de carbono; y el acero a alto car-bono tiene entre 0,60% y 1,0% de carbono.*

221

carburizing flame: A reducing oxygen-fuel gas flame in which there is an excess of fuel gas, resulting in a carbon-rich zone extending around and beyond the inner cone of the flame.

llama carburante: *Una llama reductora de oxígeno en la que hay un exceso de combus-tible, lo cual resulta en haber una zona alrededor del cono de la llama rica en carbono.*

cast iron: A family of alloys, containing more than 2% carbon and between 1% and 3% silicon. Cast irons are not malleable when solid, and most have low ductility and poor resistance to impact loading. There are four basic types of cast iron gray, white, ductile, and malleable.

hierro fundido: *Una familia de aleaciones que tienen más del 2% de carbono y entre el 1% y el 3% de silicio. Los hierros fundidos, en estado sólido, no son maleables, y la mayoría no son dúctiles y tienen poca resistencia a las cargas de impacto. Hay cuatro clases de hierro fundido, a saber: gris, blanco o especular, dúctil, y maleable.*

cathode: The negative terminal of a power supply; the electrode when using direct current electrode negative (DCEN).

cátodo: *El polo negativo de una fuente de eléctricidad. Cuando se usa el método de electrodo nega-tivo a corriente contínua, el electrodo es el cátodo.*

chain intermittent weld: An intermittent weld on both sides of a joint where the weld increments on one side are approximately opposite those on the other side.

soldadura intermitente a cadena: *Una soldadura intermitente en ambos lados de una juntura, en la que los puntos de soldadura en un lado están aproximadamente opuestos a los puntos en el otro lado.*

chill plate: A piece of metal placed behind material being welded to correct overheating.

plancha enfriadora: *Pieza de metal que se pone por detrás del material que está siendo sol-dado, para corregir el sobrecalentamiento.*

chill ring: A non-standard term for a backing ring.
anillo enfriador: Expresión, no aprobada, para anillo de soporte

chromium: A lustrous, hard, brittle, steel-gray metallic element used to harden steel alloys, in production of stainless steel, and as a corrosion resistant plating.
cromo: Elemento metálico, brillante, duro, frágil, del color del acero, que se usa para endurecer aceros de aleación, en la producción de acero inoxidable, y para hacer enchapados resistentes a la corrosión.

cladding: A surfacing variation that deposits or applies surfacing material usually to improve corrosion or heat resistance.
revestimiento: Una variación del proceso de alisamiento en la que se depositan ciertos materiales, generalmente para aumentar la resistencia a la corrosión y al calor.

coalescence: The growing together or growth into one body of the materials being welded.
coalescencia: En el caso de materiales que se están soldando, la acción de crecer juntos para formar un solo cuerpo.

coefficient of thermal expansion: The increase in length per unit length for each degree a metal is heated.
coeficiente de expansión térmica: El aumento de la longitud por cada unidad de longitud, y por cada grado de aumento de la temperatura de un metal

cohesion: Cohesion is the result of a perfect fusion and penetration when the molecules of the parent material and the added filler materials thoroughly integrate as in a weld.
adhesión: Es el resultado de una fusión y una penetración perfectas; cuando las moléculas del material soldado y las de los materiales de aportación se mezclan resueltamente, como en soldaduras.

cold crack: A crack that develops after solidification is complete.
grieta fría: La grieta que aparece después que la solidificación ha terminado.

cold work: Cold working refers to forming, bending, or hammering a metal well below the melting point. Cold working of metals causes hardening, making them stronger but less ductile.
trabajo en frío: El acto de formar, encorvar, o martillear los metales, lo cual causa un endurecimiento que los hace más fuertes pero menos dúctiles.

complete joint penetration: A root condition in a groove weld in which weld metal extends through the joint thickness.
penetración completa de la juntura: Una condición en la raíz de la juntura de una soldadura en ranura, en la cual el material de soldadura se extiende sobre todo el espesor de la juntura.

composite: A material consisting of two or more discrete materials with each material retaining its physical identity.
compuesto: Un material hecho de dos o más materiales discretos que retienen sus identidades físicas.

composite electrode: A generic term for multi-component filler metal electrodes in various physical forms such as stranded wires, tubes, or covered wire.
electrodo compuesto: Un término genérico para electrodos que contienen metales de aportación en diversas formas, v. gr., alambre retorcido, tubos, e hilos cubiertos.

concavity: The maximum distance from the face of a concave fillet weld perpendicular to a line joining the weld toes. A concave fillet weld will have a face that is contoured below a straight line between the two toes of a fillet weld.

222

concavidad: La distancia máxima desde la faz de un cordón cóncavo, perpendicular-mente, hasta la línea que toca las dos orillas de la soldadura. La superficie del cordón cóncavo está por debajo del plano que toca las dos orillas del cordón.

conductor: A device, usually a wire, used to connect or join one circuit or terminal to another.
conductor: Un dispositivo, generalmente un alambre, que sirve para conectar el terminal de un circuito al terminal de otro circuito.

cone: The conical part of an oxygen-fuel gas flame adjacent to the tip orifice.
cono: La parte cónica de una llama de oxígeno y gas combustible, adyacente al orificio de la boquilla.

constant-current (CC) power source: An arc welding power source with a volt-ampere relationship yielding a small welding current change from a large arc voltage change.
fuente de corriente contínua (CC): Un suministro de energía para soldadura con arco con una relación voltio/amperio que produce un cambio muy pequeño de corriente para un cambio grande en la tensión del arco.

constant-voltage (CV) power source: An arc welding power source with a volt-ampere relationship yielding a large welding current change from a small arc voltage change.
fuente de tensión contínua (VC): Un suministro de energía para soldadura con arco con una relación voltio/amperio que produce un cambio grande de corriente para un cambio muy pequeño en la tensión del arco.

constricted arc: A plasma arc column that is shaped by the constricting orifice in the nozzle of the plasma arc torch or plasma spraying gun.
arco restringido: Un arco de columna de plasma que surge en la forma producida por la forma de la boquilla del soplete o de la pistola del rociador de plasma.

223

A

consumable electrode: An electrode that provides filler metal, therefore is consumed in the arc welding process.
electrodo consumible: Un electrodo que es el material de aportación a la misma vez; consecuentemente, el electrodo se consume en el proceso de la soldadura con arco.

consumable insert: Filler metal that is placed at the joint root before welding, and is intended to be completely fused into the joint root to become part of the completed weld.
embutido consumible: Metal de aportación que se coloca a lo largo de la raíz de la juntura antes de empezar a soldar, para que se convierta en parte de la soldadura.

contact resistance: Resistance to the flow of electric current between two work-pieces or an electrode and the work-piece.
resistencia de contacto: La oposición al flujo de una corriente eléctrica entre dos piezas o entre un electrodo y la pieza.

convexity: The maximum distance from the face of a convex fillet weld perpendicular to a line joining the toes.
convexidad: La distancia máxima desde la faz de un cordón convexo, perpendicular-mente, hasta la línea que toca las dos orillas de la soldadura.

corner joint: A joint between two members located approximately at right angles to each other in the form of an L.
juntura de esquina: La unión de dos miembros situados aproximadamente a 90 grados uno del otro, formando una L.

covered electrode: A composite filler metal electrode consisting of a core of a bare electrode or metal cored electrode to which a covering sufficient to provide a slag layer on the weld metal has been applied. The covering may contain materials providing such functions as shielding from atmosphere, deoxidation, and arc stabilization, and can serve as a source of metallic additions to the weld.

electrodo cubierto: Un electrodo de metal de aportación compuesto que consiste de un electrodo desnudo, o un electrodo cubierto con metal, central, y el cual se recubre para que produzca una capa de escoria en el metal de soldar. Esta capa puede contener materiales para aislar la soldadura de la atmósfera, reducción, y estabilización del arco; y puede servir de fuente de añadiduras metálicas a la soldadura.

cover plate: A removable pane of colorless glass, plastics coated glass, or plastics that covers the filter plate and protects it from weld spatter, pitting, or scratching.

cubierta: Un panel removible de vidrio claro, vidrio con una capa plástica, o de plástico, que cubre el filtro y lo protege contra salpicadas, picaduras de óxido, o rasguños.

cracking a valve: Rapidly opening and closing a valve to clear the orifice of unwanted foreign material.

"abre-y-cierra" una llave: El abrir y cerrar una válvula rapidamente para limpiar el orificio de suciedades.

crater: A depression in the weld face at the termination of a weld bead.

cráter: Una depresión en la cara o parte visible de una soldadura al final del cordón de soldadura.

cutting attachment: A device for converting an oxygen-fuel gas welding torch into an oxygen-fuel cutting torch.

acesorio para cortar: Un dispositivo para cambiar un soplete de gas y oxígeno para soldar a un soplete de gas y oxígeno para cortar.

cutting head: The part of a cutting attachment to which the cutting torch or tip may be attached.

cabezal de cortar: La parte del accesorio de cortar donde se puede montar el soplete o la boquilla.

cutting tip: An attachment to an oxygen cutting torch from which the gases exit.

boquilla de cortar: Un accesorio que se monta en un soplete de cortar con oxígeno, y por donde salen los gases.

cylinder manifold: A multiple header for interconnection of gas sources with distribution points.

conector de bombonas: Una válvula de distribución para conectar diversos suministros de gas con sus puntos de uso.

D

defect: A discontinuity or discontinuities that by nature or accumulated effect render a part or product unable to meet minimum applicable acceptance standards or specifications. The term designates rejection.

defecto: Una o varias discontinuidades que por naturaleza, o por efecto cumulativo, hacen que una pieza o conjunto sea incapaz de satisfacer el mínimo estándar de aceptación o el mínimo límite de las especificaciones para las cuales fué diseñado. El término implica rechazamiento.

deposited metal: Filler metal that has been added during welding, brazing, or soldering.

metal depositado: El metal de aportación que ha sido añadido durante el proceso de soldadura.

deposition rate: The weight of filler material deposited in a unit of time.

velocidad de depósito: El peso de material depositado en la unidad de tiempo.

depth of fusion: The distance that fusion extends into the base metal or previous bead from the surface melted during welding.

profundidad de la fusión: La distancia que la fusión se extiende desde la superficie fundida hasta el metal de base o hasta un cordón de soldadura aplicado previamente.

direct current electrode negative (DCEN): The arrangement of direct current arc welding cables in which the electrode is the negative pole and the workpiece is the positive pole of the welding arc.
electrodo negativo en corriente contínua (ENCC): *El sistema de los cables del arco a corriente contínua en el que el electrodo es el polo negativo y la pieza a soldar es el polo positivo, del arco de soldadura.*

direct current electrode positive (DCEP): The arrangement of direct current arc welding cables in which the electrode is the positive pole and the workpiece is the negative pole of the welding arc.
electrodo positivo en corriente contínua (EPCC): *El sistema de los cables del arco a corriente contínua en el que el electrodo es el polo positivo y la pieza a soldar es el polo negativo, del arco de soldadura.*

distortion: Non-uniform expansion and contraction of metal caused by heating and cooling during the welding process.
deformación: *Expansiones y contracciones no uniformes producidas por el calenta-miento y enfríamiento del metal durante el proceso de soldadura.*

downhill: Welding in a downward progression.
hacia abajo: *La soldadura cuando es ejecutada de arriba hacia abajo.*

drag: During thermal cutting, the offset distance between the actual and straight line exit points of the gas stream or cutting beam measured on the exit surface of the base metal.
arrastre: *En el proceso de corte térmico, el desalineamiento que ocurre entre la línea que sigue la corriente de gas o el haz de corte en actualidad y la línea recta teórica, medido en la superficie del metal ya cortado, o en salida.*

drag angle: The travel angle when the electrode is pointing in a direction opposite to the progression of welding. This angle can also be used to partially define the positions of guns, torches, and rods.
ángulo de arrastre: *El ángulo que existe cuando el electrodo está orientado en dirección opuesta a la de la progresión de la soldadura. Este ángulo se puede usar para definir, al menos parcialmente, las posiciones de las pistolas, sopletes, varillas, y el haz de corte.*

225

ductility: The tendency to stretch or deform appreciably before fracturing.
ductilidad: *La tendencia a estirarse o deformarse mucho, antes de llegar a la fractura.*

duty cycle: The percentage of time during an arbitrary test period that a power source or its accessories can be operated at rated output without overheating. Most welding machines are rated in intervals of ten minutes meaning that a duty cycle of 50% means the machine can be operated at a given amperage setting for five continuous minutes without damage to the equipment. 60% would give six minutes; 70% would give seven minutes.
ciclo de servicio: *El porcentaje del tiempo, en un período de prueba arbitrario, en el que el suministro de energía o sus accesorios pueden ser operados al máximo valor permi-tido sin que le ocurra un sobrecalentamiento. La mayoría de las máquinas soldadoras tienen una capacidad recomendada de intervalos de diez minutos, o sea que un ciclo de servicio de 50% quiere decir que la máquina puede operar a su corriente recomendada por cinco minutos sin que la máquina sufra ningún daño; 60% daría seis minutos; 70%, siete minutos, etc.*

E

edge joint: A joint between the edges of two or more parallel or nearly parallel members.
juntura de orilla: *Una juntura entre las orillas de dos o más miembros paralelos o casi paralelos.*

edge preparation: The preparation of the edges of the joint members, by cutting, cleaning, plating, or other means.

preparación de la orilla: *La preparación de las orillas de los miembros de la juntura, cortando, puliendo, enchapando, etc. las orillas.*

effective throat: The minimum distance, minus any convexity, between the weld root and the face of a fillet weld.
garganta efectiva: *La distancia mínima (sin contar la convexidad) entre la raíz de la soldadura y la cara del cordón de soldadura.*

electrode: A component of the electrical welding circuit that terminates at the arc, molten conductive slag, or base metal.
electrodo: *Un componente del circuito eléctrico de una soldadura, que termina en el arco, la escoria fundida conductora, o el metal de base.*

electrode angle: The angle of the electrode in relationship to the surface of the material being welded; the electrode's perpendicular angle to the metals' surface leaning toward the direction of travel.
ángulo del electrodo: *El ángulo del electrodo con respecto a la superficie del material que se está soldando; el ángulo del electrodo perpendicular a la superficie del metal, inclinado hacia la dirección en que progresa la soldadura.*

electrode classification: A means of identifying electrodes by their usability, flux coverings, and chemical make up. The American Welding Society has published a series of specifications for consumables used in welding processes.
clasificación de los electrodos: *Un medio de identificar los electrodos, donde pueden usarse, fundentes, y constitución química. La American Welding Society (AWS) ha publicado una lista con especificaciones de los materiales consumibles a usar en el proceso de soldaduras.*

226

electrode holder: A device used for mechanically holding and conducting current to an electrode during welding or cutting.
boquilla de electrodo: *Un dispositivo que soporta el electrodo mecanicamente, y le trae la corriente durante la soldadura o el corte.*

electrode lead: The electrical conductor between the source of arc welding current and the electrode holder.
cables de electrodo: *El alambre eléctrico que conecta la fuente que suministra la corriente para la sodadura con arco y la boquilla que sujeta el electrodo .*

F

face reinforcement: Weld reinforcement on the side of the joint from which welding was done.
reenforzamiento de la cara: *Soldadura en el lado de la juntura por donde se había hecho la soldadura inicial para reforzar el mismo.*

faying surface: The mating surface of a member that is in contact with or in close proximity to another member to which it is to be joined.
superficie de empalme: *La superficie compañera de un miembro que está en contacto o en proximidad con otro miembro al cual se va a unir por soldadura.*

filler material: The material, metal, or alloy to be added in making a welded, brazed, or soldered joint.
material de aportación: *El material, metal, o aleación que se añade durante el proceso de soldar una juntura.*

filler metal: The metal also known as brazing filler metal, consumable insert, diffusion aid, filler material, solder, welding electrode, welding filler metal, welding rod, and welding wire.

metal de aportación: *El metal conocido también por los nombres: metal de aportación para soldaduras blandas, embutido consumible, promotor de difusión, material de aportación, soldadura, electrodo soldador, varilla soldadora, y alambre soldador.*

fillet weld: A weld of approximately triangular cross-section joining two surfaces approximately at right angles to each other in a lap joint, T-joint, or corner joint.

soldadura angular: *Un cordón de sección aproximadamente triangular, que une dos superficies colocadas a 90 grados uno del otro en una juntura a T, a empalme o de esquina.*

fillet weld leg: The distance from the joint root to the toe of the fillet weld.

pierna de la soldadura angular: *La distancia entre la raíz de la juntura y la orilla del cordón con la pieza.*

fillet weld size: For equal-leg fillet welds, the leg lengths of the largest isosceles right triangle that can be inscribed within the fillet weld cross-section. For unequal leg fillet welds, the leg lengths of the right triangle can be inscribed within the fillet weld cross-section.

tamaño de la soldadura angular: *Si las dos piernas son iguales (eso es, crean una sección en forma de triángulo isósceles), el tamaño se define como la longitud de la pierna del triángulo isósceles más grande que quepa dentro del cordón. Si las piernas son desiguales, el tamaño es la longitud de la pierna más larga del triángulo recto que se pueda inscribir en la sección.*

filter plate: An optical material that protects the eyes against excessive ultraviolet, infrared, and visible radiation. Also called filter glass or filter lens.

plato filtro: *Material óptico para proteger la vista a la exposición a demasiada radiación ultravioleta, infraroja, y visible. Tambiéen se le llama vidrio filtro o lente filtro.*

filter plate shade: Refers to the lens darkness number, which indicates the darkness of the lens.

tinte del plato filtro: *Un número que mide el poder filtrante o de absorción de ciertas radiaciones de la lente.*

227

fixture: A device designed to hold and maintain parts in proper relation to each other.

plantilla: *Un aparato que sirve para sujetar varias partes en sus posiciones relativas y fijas entre ellas.*

flame propagation rate: The speed at which flame travels through a mixture of gases.

velocidad de propagación de la llama: *La velocidad con la que viaja una llama por una mezcla de gases.*

flare-V-groove weld: A weld in a groove formed by two members with curved surfaces.

soldadura de ranura en V ensanchada: *Soldadura hecha en una ranura formada por dos miembros con superficies curvas.*

flashback: A recession of the flame into or back of the mixing chamber of the oxygen fuel gas torch or flame spraying gun.

contraquema: *La reversión de la llama hacia la cámara donde se mezcla el gas combustiblle con el oxígeno en sopletes y en pistolas rociadoras de llama.*

flashback arrester: A device to limit damage from a flashback by preventing propagation of the flame from beyond the location of the arrester.

arrestador de contraquema: *Un aparato que sirve para reducir el daño causado por la retroquema, evitando que la llama viaje hacia atrás.*

flat welding position: The welding position used to weld from the upper side of the joint at a point where the weld axis is approximately horizontal, and the weld face lies in an approximately horizontal plane.

posición de soldar plana: *La posición que se usa para soldar desde la parte alta de la juntura, en un punto donde el eje de la soldadura es aproximadamente horizontal y la cara del cordón se halla en un plano aproximadamente horizontal.*

flaw: An undesirable blemish or discontinuity in a weld such as a crack or porosity.
defecto: *Una imperfección indeseable o discontinuidad como una grieta o porosidad.*

flux: A material used to hinder or prevent the formation of oxides and other undesirable substances in molten metal and on solid metal surfaces, and to dissolve or otherwise facilitate the removal of such substances.
fundente: *Un material que sirve para retrasar o prevenir la formación de óxidos y otras substancias indeseables en el metal fundido y en las superficies de la juntura; y para disolver o ayudar en la remoción de esas substancias.*

flux cored electrode: A composite tubular filler metal electrode consisting of a metal sheath and a core of various powdered materials producing an extensive slag cover on the face of a weld bead. External shielding may be required.
electrodo con núcleo de fundente: *Un electrodo con metal de aportación, tubular compuesto, que consiste de una envoltura cilíndrica de metal y, en el centro, varios materiales pulverizados que producen una capa de escoria que protege la cara del cordón de soldadura. Es posible que se requiera protección externa.*

forehand welding: A welding technique in which the welding torch or gun is directed toward the progress of welding.
soldando con la palma: *Una técnica de soldar en la cual se hace que la boquilla del soplete esté orientada en la misma dirección que aquella en la que progresa la soldadura.*

228

fuel gas: A gas such as acetylene, natural gas, hydrogen, propane, stabilized methylacetylene propadiene, and other fuels normally used with oxygen in one of the oxyfuel processes and for heating.
gas combustible: *Un gas como acetileno, gas natural, hidrógeno, propano, metil-acetileno propadieno, y otros, usado normalmente con oxígeno en los procesos llamados en inglés "oxyfuel," y para calentar.*

fusible plug: A metal alloy plug that closes the discharge channel of a gas cylinder and is designed to melt at a predetermined temperature permitting the escape of gas.
tapón fusible: *Un tapón hecho de metal de aleación que tapa la salida del gas de una bombona, y no lo deja salir hasta que haya llegado a cierta temperatura.*

fusion: The joining of base material, with or without filler material, by melting them together.
fusión: *La unión de materiales de base, con o sin material de aportación. Esta unión se hace fundiendo las piezas juntas.*

fusion face: A surface of the base metal that will be melted during welding.
cara de fusión: *Una superficie del metal de base, que será fundida en la soldadura.*

fusion welding: Any welding process that uses fusion of the base metal to make the weld.
soldadura a fusión: *Cualquier proceso en el que la soldadura se obtiene fundiendo el metal de base.*

fusion zone: The area of base metal as determined on the cross-section of a weld.
zona de fusión: *El área del metal de base determinada analizando la sección transversal de la soldadura.*

G

gas cylinder: A portable container used for transportation and storage of compressed gas.
bombona (o cilindro) de gas: *Un recipiente portátil para la transportación y el almacenamiento de gas comprimido.*

gas nozzle: A device at the exit end of the torch or gun that directs shielding gas.
boquilla de gas: Un dispositivo montado a la salida del soplete o pistola que dirije el gas protector.

gas regulator: A device for controlling the delivery of gas at some substantially constant pressure.
regulador de gas: Un dispositivo para controlar que la salida del gas ocurra a una presión más o menos constante.

globular transfer: In arc welding, the transfer of molten metal in large drops from a consumable electrode across the arc.
transferencia globular: En soldaduras con arco, el traspaso de metal fundido en grandes gotas, del electrodo consumible através del arco.

GMAW: The welding process Gas Metal Arc Welding; non-standard terms for this process are MIG (metal inert gas), MAG (metal active gas), wire feed, hard wire welding.
GMAW: Sigla americana para denotar Soldadura con Arco usando Metal y Gas. Otros nom bres (no aprobados) son: MIG (Metal y Gas Inerte), MAG (Metal y Gas Activo), alimen tado con alambre, y soldadura con alambre duro.

goggles: Protective glasses equipped with filter plates set in a frame that fits snugly against the face and used primarily with oxygen fuel gas welding processes.
anteojos de seguridad: Anteojos equipados con lentes filtrantes montados en una montura que se pega muy bien en la cara, para proteger los ojos. Se usan mayormente cuando se está soldando con oxígeno/combustible.

groove angle: The total included angle of the groove between workpieces.
ángulo de ranura: El ángulo formado por la ranura que aparece al juntar dos piezas para hacer una soldadura.

229

groove face: The surface of a joint member included in the side of the groove from root to toe.
cara de la ranura: La superficie de un miembro de una juntura incluído en el lado de la ranura, desde la raíz hasta la orilla.

groove radius: The radius used to form the shape of a J- or U-groove weld.
radio de la ranura: El radio usado para hacer una soldadura de ranura en la forma de una J o de una U.

groove weld: A weld made in a groove between the workpieces. See welding symbols.
soldadura de ranura: La soldadura formada en la ranura creada entre dos piezas. Véanse los símbolos de soldadura en el texto.

groove weld size: The joint penetration of a groove weld. Also groove throat or effective throat.
tamaño de la soldadura de ranura: En una soldadura de ranura, la penetración de la juntura. También se le llama garganta de la soldadura, o garganta efectiva.

ground connection: An electrical connection of the welding machine frame to the earth for safety.
Conexión a tierra: Una conexión eléctrica de la máquina soldadora a la tierra, para seguridad.

GTAW: The Gas Tungsten Arc Welding process; non-standard terms are Heliarc™, and TIG (tungsten inert gas).
GTAW: Sigla americana que denota soldadura con arco usando tungsteno y un gas inerte. Otros nombres (no aprobados) son: heliarc y TIG (tungsteno y gas inerte).

H

hardfacing: A surfacing variation in which hard material is deposited to reduce wear.
revestimiento duro: Una variación de revestimiento en la que material es depositado en la superficie para reducir el desgaste.

heat-affected zone (HAZ): The portion of the base metal whose mechanical properties or microstructure have been altered by the heat of welding, brazing, soldering, or thermal cutting.
zona afectada por el calor: Aquella porción del metal de base cuyas propiedades mecá-nicas o su micro-estructura han sido alteradas por el calor creado al soldar o al cortar termicamente.

high carbon steel: See carbon steel.
acero a alto crbono: Véase debajo de la definición en inglés de "carbon steel."

horizontal welding position: In a fillet weld, the welding position in which the weld is on the upper side of an approximately horizontal surface and against an approximately vertical surface. In a groove weld, the welding position in which the weld face lies in an approximately vertical plane and the weld axis at the point of welding is approximately horizontal.
posición de soldadura horizontal: En una soldadura angular, la posición de soldar en la que el cordón se aplica en el lado de arriba de una superficie aproximadamente horizontal, y contra otra superficie aproximadamente vertical. En la sodadura de una ranura, la posición en la que la cara de la soldadura está contenida en un plano aproximadamente vertical, y el eje de la soldadura al punto de la soldadura es aproximadamente horizontal.

I

impact strength: The ability of a material to resist shock, dependent on both strength and ductility of the material.
resistencia al impacto: La habilidad de un material a resistir una carga de choque y que depende de su resistencia y su ductilidad.

inclusion: Entrapped foreign solid material, such as slag, flux, tungsten, or oxide.
inclusión: Partícula extraña de metal sólida, así como escoria, fundente, tungsteno, u óxidos, que se halla atrapada en la soldadura.

inert gas: A gas that normally does not combine chemically with other elements or compounds.
gas inerte: Un gas que normalmente no se combina quimicamente con ningún otro material.

intermittent weld: A weld in which the continuity is broken by recurring unwelded spaces.
soldadura intermitente: Una soldadura en la cual la continuidad está interrumpida por espacios que no están soldados.

interpass temperature: In a multipass weld, the temperature of the weld area between weld passes.
temperatura entre pases: En una soldadura con muchos pases, se refiere a la temperatura de la zona de soldadura entre pases.

inverter power supply: A welding power supply with solid-state electrical components that change the incoming 60 Hz power to a higher frequency. Changing the frequency results in greatly reducing the size and weight of the transformer. Inverter power supplies can be used with all arc welding processes.
alimentador eléctrico a inversión: Un suministrador de enrgía para soldar, con componentes eléctronicos de estado sólido que convierten la entrada de corriente de 60 Hertz a una frequencia muy alta, para poder usar transformadores más pequeños y más ligeros. Este tipo de alimentador puede usarse en todos los procesos para soldar.

iron carbide: A binary compound of carbon and iron; it becomes the strengthening consitituent in steel.
carburo de hierro: Compuesto binario de carbono que contiene más elementos electropositivos, y que en combinación con el hierro, es el constituyente que da la resistencia al acero.

iron soldering: A soldering process in which the heat required is obtained from a soldering iron.

soldadura con hierro de soldar: Un proceso de soldadura en el que el calor para soldar se obtiene de un hierro de soldar.

J

J-groove weld: A type of groove weld where one side of the joint forms a J.
soldadura en ranura en J: Un tipo de soldadura en el que un lado de la juntura tiene la forma de una jota mayúscula.

joint: The junction of members or the edges of members that are to be joined or have been joined.
juntura: La unión de dos miembros, o las orillas de los miembros que van a ser soldados.

joint clearance: The distance between the faying surfaces of a joint in brazing or soldering.
juego en la juntura: La distancia entre las superficies a unirse en soldaduras fuertes o blandas.

joint design: The shape, dimensions, and configuration of the joint.
diseño de una juntura: La forma, dimensiones y configuración de una juntura.

joint efficiency: The ratio of strength of a joint to the strength of the base metal expressed in percent.
eficacia de una juntura: la razón de la resistencia mecánica de una juntura a la resistencia del metal de base.

joint filler: A metal plate inserted between the splice member and thinner joint member to accommodate joint members of dissimilar thickness in a spliced butt joint.
relleno de juntura: En una juntura de tope a empalme, el plato que se introduce entre el empalme y el miembro más delgado, para acomodarlo con el miembro mas grueso.

joint geometry: The shape and dimensions of a joint in cross-section prior to welding.
geometría de la juntura: La forma y las dimensiones de la sección de una juntura antes de ser soldada.

231

joint penetration: The distance the weld metal extends from the weld face into a joint, exclusive of weld reinforcement.
penetración de la juntura: la distancia que el material de soldadura se ha extendido de la cara del cordón hasta el interior de la juntura, sin incluir ninguna soldadura de refuerzo.

joint root: That portion of a joint to be welded where the members approach closest to each other. In cross-section, the joint root may be a point, a line, or an area.
raíz de la juntura: Aquella porción de la juntura a soldarse, donde los miembros están más cerca uno al otro. Viéndola en sección, la raíz de una juntura puede ser un punto, o una línea, o un plano.

joint spacer: A metal part, such as a strip, bar, or ring, inserted in the joint root to serve as a backing and to maintain the root opening during welding.
separador de juntura: Una pieza metálica, v. gr. tira delgada de relleno, o una barra, o un anillo, la cual causa que la raíz esté abierta durante la soldadura.

joint type: A weld joint classification based on five basic joint configurations such as a butt joint, corner joint, edge joint, lap joint, and T-joint.
tipo de juntura: Clasificación de junturas basada en cinco configuraciones, a saber: juntura de tope, juntura de esquina, juntura de orilla, juntura de empalme, y juntura en forma de T.

K

kerf: The width of a cut produced during a cutting process.
entalladura: El espesor del corte producido durante el corte.

keyhole welding: A technique in which a concentrated heat source penetrates partially or completely through a workpiece, forming a hole (or keyhole) at the leading edge of the weld pool. As the heat source progresses, the molten metal fills in behind the hole to form the weld bead.

soldadura a bocallave: *Una ejecución en la que un aparato generador de calor se introduce, parcial or totalmente, através de la pieza creando un hueco (de ahí el nombre de bocallave) en la orilla de la balsa de soldadura. A medida que la introducción del calentador progresa, el metal líquido se cuela por el hueco, formando un cordón.*

L

lamination: A type of discontinuity with separation or weakness generally aligned parallel to the worked surface of a metal.

laminación: *Un tipo de discontinuidad con separaciones o debilidad normalmente alineadas en un plano paralelo a la superficie externa que esta siendo labrada.*

lap joint: A joint between two overlapping members in parallel planes.

juntura a empalme: *Una juntura formada por dos miembros asolapados en planos paralelos.*

laser beam cutting (LBC): A thermal cutting process that severs metal by locally melting or vaporizing with the heat from a laser beam.

corte con rayo laser: *Un proceso térmico que separa material fundiendo o vaporizando el metal localmente con el calor de un rayo laser.*

lens shade: See filter plate shade.

sombra de lente: *Véase debajo de filter plate shade arriba.*

liquidus: The lowest temperature at which a metal or an alloy is completely liquid.

liquidus: *La mínima temperatura a la que un metal o una aleación se mantiene líquida.*

longitudinal crack: A crack with its major axis orientation approximately parallel to the weld axis.

grieta longitudinal: *Una grieta cuyo eje mayor está orientado aproximadamente paralelo al eje del cordón de soldadura.*

M

manifold: See cylinder manifold.

connector: *Véase debajo de cylinder manifold arriba.*

manual welding: Welding with the torch, gun, or electrode holder held and manipulated by hand. Accessory equipment, such as part motion devices and manually controlled filler material feeders may be used.

soldadura manual: *Soldadura hecha teniendo el soplete, o la boquilla del electrodo en la mano. Accesorios como dispositivos para mover la pieza, alimentadores de material de aportación operados manualmente son permitidos.*

medium steel: Refer to carbon steel.

acero mediano en carbono: *Refiérase a debajo de carbon steel arriba.*

melt-through: Visible root reinforcement produced in a joint welded from one side.

fundido completo: *Refuerzo visible en la raíz, producido por la soldadura de una juntura en un lado nada más.*

metal-cored electrode: A composite tubular filler metal electrode consisting of a metal sheath and a core of various powdered materials.

electrodo con núcleo de metal: *Un electrodo tubular compuesto, con metal de aportación, que consiste de una vaina metálica y un núcleo de varios materiales pulverizados.*

metal electrode: A filler or non-filler metal electrode used in arc welding or cutting, which consists of a metal wire or rod that has been manufactured by any method and that is either bare or covered with a suitable covering or coating.
electrodo de metal: *Un electrodo de metal — con o sin metal de aportación – usado en corte y soldadura con arco, que consiste de un alambre o varilla metálica, construído por cualquier método, y que está desnudo o cubierto apropiadamente.*

metallic bond: The principal atomic bond that holds metals together.
afinidad metálica: *El enlace principal que mantiene a los metales enteros.*

metallurgy: The science explaining the properties, behavior, and internal structure of metals.
metalurgia: *La ciencia que trata de las propiedades, el comportamiento, y la estructura interna de los metales.*

methylacetylene propadiene: A family of alternative fuel gases that are mixtures of two or more gases (propane, butane, butadiene, methylacetylene, and propadiene). Methylacetylene propadiene is used for oxyfuel cutting, heating, brazing, and soldering.
metilacetileno propadieno: *Una familia de gases combustibles que son mezclas de dos o más gases (propano, butano, butadieno, metilacetileno, y propadieno).*

mild steel: Refer to carbon steel.
acero dulce: *Refiérase a debajo de carbon steel arriba.*

mixing chamber: That part of a welding or cutting torch in which a fuel gas and oxygen are mixed.
región de mezcla: *La parte del soplete de cortar o soldar en donde el gas combustible y el oxígeno se mezclan.*

multipass welding: A weld requiring more than one pass to ensure complete and satisfactory joining of the metal pieces.
soldadura de varios pases: *Una soldadura que requiere más de un pase para asegurar la unión satisfactoria de las piezas metálicas.*

N

neutral flame: An oxyfuel gas flame that has characteristics neither oxidizing nor reducing.
llama neutra: *Una llama de oxígeno + gas combustible que no tiene tendencias ni a oxidar ni a reducir.*

nitrogen: A gaseous element that occurs freely in nature and constitutes about 78% of earth's atmosphere. Colorless, odorless, and relatively inert, although it combines directly with magnesium, lithium, and calcium when heated with them. Produced either by liquefaction and fractional distillation of air, or by heating a water solution of ammonium nitrate.
nitrógeno: *Elemento gaseoso muy común en la naturaleza terrestre: representa el 78% de la atmósfera que circunda el planeta. Es incoloro, inodoro y relativamente inerte. Sin embargo, se combina facilmente con magnesio, el litio, y calcio en la presencia de calor.*
Se obtiene por licuefacción, seguida por destilación fraccional, del aire; o calentando una solución ácuea de nitrato de amonio.

non-consumable electrode: An electrode that does not provide filler metal, as used in the GTAW process.
electrode no consumible: *Un electrodo que no provee el metal de aportación. Un ejemplo es el electrodo para corte con arco de carbono.*

non-corrosive flux: A soldering flux that in either its original or residual form does not chemically attack the base metal. It usually is composed of rosin-based materials.

233

fundente no corrosivo: Un fundente que no combina con el metal de base, en su forma original o su forma residual. Normalmente deriva de materiales resinosos.

normalizing: The process of heating a metal above a critical temperature and allowing it to cool slowly under room temperature conditions to obtain a softer and less distorted material.
normalizar: El proceso de calentar un metal por encima de su temperatura crítica, y dejar que se enfríe lentamente a la temperatura del laboratorio, para obtener un material que es más blando y menos deformado.

O

ohm: A unit of electrical resistance. An ohm is equal to resistance of a circuit in which a potential difference of one volt produces a current of one ampere.
ohmio: La unidad de resistencia eléctrica. Un ohmio es la resistencia de un circuito en el que una diferencia de potencial de un voltio, genera una corriente de un amperio

open-circuit voltage: The voltage between the output terminals of the power source when no current is flowing to the torch or gun.
tensión con circuito abierto: El voltaje en los terminales de salida del alimentador de potencia cuando no hay ninguna corriente pasando por la pistola o el soplete.

open root joint: An unwelded joint without backing or consumable insert.
juntura con raíz abierta: Una juntura no soldada, sin soporte ni embutido de consumibles.

overlap: The protrusion of weld metal beyond the weld toe or weld root.
asolapado: El metal de la soldadura que sobresale más allá de la orilla con el metal de base, ode la raíz de la soldadura.

oxidizing flame: An oxyfuel flame in which there is an excess of oxygen, resulting in an oxygen-rich zone extending around and beyond the cone.
llama oxidante: Una llama de oxígeno/gas combustible en la que hay exceso de oxígeno, lo cual resulta en haber una zona alrededor y cerca del cono rica en oxígeno.

oxygen: A colorless, odorless, tasteless, gaseous chemical element, the most abundant of all elements. Oxygen occurs free in the atmosphere, forming 1/5 of its volume, and in combination in water, sandstone, limestone, etc.; it is very active being able to combine with nearly all other elements and is essential to life.
oxígeno: Elemento químico, incoloro, inodoro, insípido, gaseoso; el elemento más abundante en la Tierra. Se encuentra, libre, en la atmósfera, de la que forma la quinta parte de su volúmen. También se halla, en combinación, en el agua, piedra de arena, piedra de cal, etc. Es muy activo, y tiende a combinarse con practicamente todos los demás elementos. Por último, es indispensable para la vida.

P

parent metal: A non-standard term referring to the base metal.
metal padre: Un término, no aprobado, para referirse al metal de base.

partial joint penetration weld: A joint root condition in a groove weld in which incomplete joint penetration exists.
soldadura con penetración parcial de la juntura: En una soldadura de ranura, la condición en la raíz de la juntura en la que existe una penetración parcial de la soldadura.

pass: A single progression of welding along a joint, resulting in a weld bead or layer.
pase: Una sola progresión de soldadura a lo largo de una juntura, lo que resulta en una capa o cordón.

peening: The mechanical working of metals using impact blows.

martilleado: El trabajo mecánico hecho en una pieza metálica por medio de una serie seguida de martillazos.

penetration: A non-standard term used in describing depth of fusion, joint penetration, or root penetration.

penetración: Un término, no aprobado, para describir la profundidad de una fusión, penetración de una juntura, o penetración de la raíz.

phase transitions: When metals or metal alloys go from solid to liquid or the reverse. These changes are also called allotropic transformations.

transiciones de fase: Una transición de fase ocurre, por ejemplo, cuando un metal de aleación pasa de su fase sólida a su fase líquida, o vice versa. Estas transiciones se llaman también transformaciones alotrópicas.

pilot arc: A low-current arc between the electrode and the constricting nozzle of the plasma arc torch to ionize the gas and facilitate the start of the welding arc.

arco piloto: Un arco a baja corriente entre el electrodo y la boquilla constrictiva en el soplete con arco de plasma para ionizar el gas, y así facilitar el inicio del arco de sol dadura.

plasma arc cutting (PAC): An arc cutting process that uses a constricted arc and removes the molten metal with a high-velocity jet of ionized gas issuing from the constricting orifice.

corte con arco de plasma: Un proceso de corte con arco que usa un arco constringido y remueve el metal fundido con un chorro a alta velocidad de gas ionizado que sale por el orificio constringido.

235

plug weld: A weld made in a circular hole in one member of a joint fusing that member to another member.

soldadura de tapón: Una soldadura en un orificio circular de un miembro de una juntura fundiendo ese miembro con el otro.

polarity: The condition of being positive or negative with respect to some reference point or object. In welding the terminals of the power supply are designated negative and positive. Whichever terminal is hooked to the electrode determines polarity.

polaridad: La condición de ser positivo o negativo con respecto a un punto u objeto de referencia. En las máquinas para soldadura, los terminales del alimentador de energía están marcados positivo y negativo. El terminal al que se conecta el electrodo determina la polaridad.

positive pressure torch: The positive pressure torch requires that gases be delivered at pressures above 2 psi (14 kPa). In the case of acetylene, the pressure should be between 2 and 15 psi (14 to 103 kPa). Oxygen is generally supplied at approximately the same pressure for welding.

antorcha a presión: Un soplete el cual se debe alimentar con gases a presiones de 2 psi (14 kPa) o mayor. En el caso del acetileno, la presión debe ser entre psi (14 kPa) y 15 psi (103 kPa). El oxígeno también se suministra a estas presiones para soldaduras.

power source: An apparatus for supplying current and voltage suitable for welding, thermal cutting, or thermal spraying.

fuente de energía: Un aparato para el suministro de la corriente y tensión apropiadas para soldar, cortar termicamente o rociar termicamente.

precipitate: To cause to become insoluble, with heat or a chemical reagent, and separate out from a solution.

precipitar: Causar a una substancia en solución que se vuelva insoluble usando calor o un reactivo químico, y que se deposite en forma sólida.

preform: Brazing or soldering filler metal fabricated in a shape or form for a specific application.

preforma: La forma que se le da a un metal de aportación en su fabricación para su uso en un caso particular.

preheat: The heat applied to the base metal or substrate to attain and maintain preheat temperature.

pre-calentamiento: El calor aplicado a un metal de base o substrato para obtener y mantener una temperatura de calentamiento predeterminada.

pressure regulator: A device designed to maintain a nearly constant supply pressure. Regulators may be attached to pressurized cylinders, gas generators, or pipe lines to reduce pressure as desired to operate equipment.

regulador de presión: Un dispositivo para mantener el suministro de una presión casi con-stante. Los reguladores pueden ser conectados a bombonas de gas comprimido, generadores a gas, o líneas de tuberías para reducir la presión al valor necesario para operar la maquinaria.

protective atmosphere: A gas or vacuum envelope surrounding the workpieces, used to prevent or reduce the formation of oxides and other detrimental surface substances, and to facilitate their removal.

atmósfera protectora: Una zona de gas o de vacío que envuelve las piezas a soldarse, y se usa para evitar o reducir la formación de óxidos y otras substancias de detrimento a sus superficies; y para facilitar su remoción.

pulsed-power welding: An arc welding process variation in which the power is cyclically programmed to pulse so that effective but short duration values of power can be utilized. Such short duration values are significantly different from the average value of power. Equivalent terms are pulsed-voltage or pulsed-current welding.

soldadura a pulsos: Una variación del proceso de soldadura con arco en la cual la energía es suministrada en pulsos a intervalos de corta duración pero muy potentes. Estos pulsos son mucho más altos que la potencia media. Otros términos son: soldadura a voltage pulsado o soldadura a corriente pulsada.

purging: The removing of any unwanted gas or vapor from a container, chamber, hose, torch, or furnace.

purgar: La remoción de gases o vapores indeseables de un recipiente, cámara, mangas, sopletes u hornos.

push angle: The travel angle when the electrode is pointing in the direction of the weld progression. This angle can also be used to partially define the positions of welding guns.

ángulo de empuje: El ángulo de curso cuando el electrodo está inclinado en la misma dirección del progreso de la soldadura. Este ángulo se puede usar para definir, en parte, la posición de los sopletes.

Q

quenching: The sudden cooling of heated metal by immersion in water, oil, or other liquid. The purpose of quenching is to produce desired strength properties in hardenable steel.

temple: En un metal muy caliente, el enfriamiento rápido obtenido por inmersión del metal en un baño de agua, aceite, u otro líquido. El propósito es el de obtener ciertas características en las propiedades de los aceros endurecibles.

R

reactor: A device used in arc welding circuits to minimize or smooth irregularities in the flow of the welding current; also called an inductor.

reactancia: Un dispositivo eléctrico que se usa en soldaduras con arco para eliminar o minimizar fluctuaciones de la corriente del arco. También se le llama inductancia.

reducing flame: An oxyfuel flame with an excess of fuel gas.
llama reductora: una llama de oxígeno/gas combustible con exceso de combustible.

resistor: A device with measurable, controllable, or known electrical resistance used in electronic circuits or in arc welding circuits to regulate the arc amperes.
resistor: Un componente eléctrico con una resistencia elétrica que puede ser medida, controlada, y conocida. Se usa en circuitos eléctricos; y en circuitos de soldaduras con arco, sirve para controlar el amperaje del arco.

root bead: A weld bead that extends into or includes part or all of the joint root.
cordón de raíz: Un cordón de soldadura que se extiende dentro de la raíz de la juntura incluyendo parte de, o toda ella.

root face: That portion of the groove face within the joint root.
ccara de la raíz: La porción de la cara de la ranura que se encuentra dentro de la raíz de la juntura.

root opening: A separation at the joint root between the workpieces.
apertura de la raíz: El espacio, en la raíz de la juntura, entre los miembros de la juntura.

root penetration: The distance the weld metal extends into the joint root.
penetración de la raíz: La distancia que el metal de soldadura se extiende dentro de la raíz de la juntura.

root reinforcement: Weld reinforcement opposite the side from which welding was done.
refuerzo de la raíz: Cordón de refuerzo en el lado opuesto al de la soldadura original.

runoff weld tab: Additional material that extends beyond the end of the joint, on which the weld is terminated.
pestaña del fin the soldadura: Exceso de material que se extiende más allá del fin de la juntura, y donde la soldadura termina.

S

safety disc: A disc in the back side of a high pressure cylinder valve designed to rupture and release gas to the atmosphere preventing cylinder rupture if the cylinder is mishandled.
disco de seguridad: En una bombona de gas comprimido, un disco en la parte trasera de la válvula de la bombona hecho de modo que se rompa y deje escapar el gas más bien que dejar que la bombona explote por mal tratamiento.

seal weld: Any weld designed primarily to provide a specific degree of tightness against leakage.
cordón de sello: Un cordón diseñado para producir cierto grado de impermeabilidad.

seam weld: A continuous weld made between or upon overlapping members, in which coalescence may start and occur on the faying surfaces, or may have proceeded from the outer surface of one member. The continuous weld may consist of a single weld bead or a series of overlapping spot welds.
soldadura de costura: Una soldadura contínua hecha entre, o por encima de miembros de una juntura asolapados, y en las que coalescencia puede empezar a ocurrir en las superficies de la juntura, o puede que venga de la superficie externa de uno de los miembros. La soldadura contínua puede consistir de un cordón simple, o de una serie de soldaduras por puntos.

shear: To tear or wrench by shearing stress; to cut through using a cold cutting tool when shearing metal.
romper con fuerza cortante: Rasgar o deslocar por medio de esfuerzo cortante. Cuando se trata de metales, cortar a través de una pieza usando una herramienta para cortar en frío.

237

shielding gas: Protective gas used to prevent or reduce atmospheric contamination.
gas aislante: Gas protector, que elimina, o reduce, la contaminación atmosférica.

short-circuiting transfer: Metal transfer in which molten metal from a consumable electrode is deposited during repeated short circuits.
transferencia con corto-circuitos: El caso en que metal fundido de un electrodo consumible es depositado durante corto-circuitos repetidos.

silicon: A non-metallic element resembling graphite in appearance, used extensively in alloys. It is the second most common element on earth. Silicon is usually found in the oxide (silicate) form. Silicon contributes to the strength of low-alloy steels and increases hardenability along with performing the valuable function of a deoxidizer, eliminating trapped gas.
silicio: Elemento no metálico, muy parecido al grafito, muy usado como elemento de aleación. Se encuentra como óxido (silicato). El silicio contribuye a mejorar la resistencia de aceros de baja aleación, y aumenta su cementación; y a la misma vez desempeña la importante función de desoxigenante, eliminando los gases que están atrapados en el metal.

silicon rectifier: A silicon semiconductor device that acts like a check valve for electricity and is used to change alternating current to direct current.
rectificador de silicio: Un dispositivo hecho del elemento semiconductor silicio, que funciona como una válvula unidireccional para la electricidad, por lo que se usa para convertir una corriente alterna en corriente contínua.

single-phase: A generator or circuit in which only one alternating current voltage is produced.
monofásico: Un generador o circuito en el que se produce una sola corriente o voltaje alterno.

slag: A nonmetallic product resulting from the mutual dissolution of flux and non-metallic impurities in some welding and brazing processes.
escoria: El producto, no metálico, de la disolución del fundente y de impurezas no metá-licas que ocurre en algunos tipos de soldadura.

slope: A term used to describe the shape of the static volt-ampere curve of a constant-voltage welding machine. Slope is caused by impedance and is usually introduced by adding substantial amounts of inductance to the welding power circuit. As more inductance is added to a welding circuit, there is a steeper slope to the volt-ampere curve. The added inductance limits the available short-circuit current and slows the rate of response of the welding machine to changing arc conditions.
inclinación: Término que describe la forma de la curva estática de volt-amperios en una máquina soldadora a voltaje constante. La inclinación es causada por la impedancia que resulta del aumento de inductancia en el circuito que suministra la energía. A más inductancia, más tiende la curva a ser vertical. Añadiendo inductancia limita la corriente en corto-circuitos y retarda la reacción de la máquina a cambios en la condición del arco.

slot weld: A weld made in an elongated hole in one member of a joint fusing that member to another member. The hole may be open at one end.
soldadura de ranura: Soldadura hecha en un orificio alargado en un miembro de la juntura, fundiéndolo sobre otro miembro. El orificio puede estar abierto en un extremo.

slugging: The unauthorized addition of metal, such as a length of rod, to a joint before welding or between passes, often resulting in a weld with incomplete fusion.
aporreando: La adición, no aprobada, de metal, como por ejemplo un pedazo de varilla, en una juntura antes de soldarla, o entre pases, y que resulta en una soldadura con fusión incompleta.

solder: The metal or alloy used as a filler metal in soldering, which has a liquidus not exceeding 840°F (450°C) and below the solidus of the base metal.

soldadura: El metal o aleación que se usa como material de aportación en soldaduras blandas, cuyo liquidus no sobrepasa los 840°F (450°C) y es menor que el liquidus del metal de base.

soldering: A group of welding processes that produce coalescence of materials by heating them to the soldering temperature and by using a filler metal having a liquidus not exceeding 840°F (450°C) and below the solidus of the base metals. The filler metal is distributed between closely fitted faying surfaces of the joint by capillary action.

soldar: Un grupo de procesos de soldadura que produce coaleescencia de materiales calentándolos hasta llegar a la temperatura de soldar y usando un metal de aportación cuyo liquidus no sobrepasa los 840°F (450°C) y es menor que el liquidus de los materiales de base. El metal de aportación se distribuye por todas las superficies de la juntura en contacto íntimo por efecto de capilaridad.

soldering iron: A soldering tool having an internally or externally heated metal bit usually made of copper.

hierro de soldar: Un utensilio que tiene una punta metálica calentada internamente o externamente , que muchas veces es hecha de cobre.

solder interface: The interface between solder metal and the base metal in a soldered joint.

superficies colindantes de una soldadura: el espacio entre el metal de soldadura y el metal de base. Se le llama in-ter-féis en inglés.

solder metal: That portion of a soldered joint that has melted during soldering.

metal de soldar: Aquella porción de la juntura soldada que se fundió durante el proceso de soldarla.

solidus: The highest temperature at which a metal or an alloy is completely solid.

sólidus: En un metal o en una aleación, la máxima temperature a la que el material está todavía completamente sólido.

239

spacer strip: A metal strip or bar prepared for a groove weld and inserted in the joint root to serve as a backing and to maintain the root opening during welding. It can also bridge an exceptionally wide root opening due to poor fit.

tira de separación: Una tira o barra de metal preparada para montarse en la raíz de la juntura de una soldadura de ranura. Sirve de soporte y también mantiene la raíz abierta durante la soldadura. Puede ser montada en casos de raíces muy amplias debido a una juntura mal montada con mucho juego.

spatter: The metal particles expelled during fusion welding that do not form a part of the weld.

salpicadura: Partículas metálicas lanzadas durante la soldadura a fusión, y que no forman parte de la soldadura.

spliced joint: A joint in which an additional workpiece spans the joint and is welded to each member.

juntura empalmada: Juntura en la que se monta otra pieza que cubre la juntura y sus bordes se sueldan a cada miembro.

spool: A filler metal package consisting of a continuous length of welding wire in coil form wound on a cylinder (called a barrel) which is flanged at both ends. The flange contains a spindle hole of smaller diameter than the inside diameter of the barrel.

bobina: Un paquete de metal de aportación en forma de alambre enrollado sobre un cilindro (llamado barril) con rebordes en los lados. El reborde tiene un orificio cuyo diámetro es menor que el del barril.

spot weld: A weld made between or upon overlapping members in which coalescence may start and occur on the faying surfaces or may proceed from the outer surface of one member. The weld cross-section is approximately circular.

soldadura por puntos: *Una soldadura hecha a una juntura asolapada en la que coalescencia puede empezar en las superficies de la juntura, o en la superficie externa de uno de los miembros. La sección de la soldadura es aproximadamente circular.*

spray transfer: Metal transfer in which molten metal from a consumable electrode is propelled axially across the arc in small droplets.
traslado con rocío: *Trnsferencia de metal en el que el metal fundido proviene del electrodo consumible, y es impelido axialmente a través del arco en la forma de gotas pequeñas.*

staggered intermittent weld: An intermittent weld on both sides of a joint with the weld increments on one side alternating with respect to those on the other side.
soldadura escalonada intermitente: *Una soldadura intermitente de ambos lados de una juntura y con los puntos de soldadura de un lado intercalados entre los puntos del otro lado.*

standoff distance: The distance between a welding nozzle and the workpiece.
distancia de compensación: *La distancia de la boquilla del soplete a la pieza.*

steel: A material composed primarily of iron, less than 2% carbon, and (in an alloy steel) small percentages of other alloying elements.
acero: *Material compuesto principalmente de hierro, menos de 2% de carbono, y (en los aceros de aleación) porcentajes pequeños de otros elementos de aleación.*

step-down transformer: A transformer that reduces the incoming voltage.
transformador reductor: *Un transformador que reduce la (alta) tension, o corriente de entrada en un valor bajo de las mismas.*

step-up transformer: A transformer that increases the incoming voltage.
transformador elevador: *Un transformador que eleva la (baja) tension, o corriente de entrada en un valor alto de las mismas.*

stickout: In GTAW, the length of the tungsten electrode extending beyond the end of the gas nozzle. In GMAW and FCAW, the length of the unmelted electrode extending beyond the end of the contact tube.
extensión: *Usando GTAW, la distancia que el electrodo de tungsteno se extiende después de pasar la boquilla de gas. Usando GMAW y FCAW, la longitud del electrodo que se extiende después de pasar el tubo de contacto.*

strain: Distortion or deformation of a metal structure due to stress.
deformación: *Distorsión o deformación de la estructura metálica debida a los esfuerzos.*

stress: A force causing or tending to cause deformation in metal. A stress causes strain.
esfuerzo: *Condición interna de una substancia elástica causada por cargas externas, y que va acompañada por una codición de deformaciones en la substancia.*

stringer bead: A type of weld bead made without appreciable weaving motion.
cordón de zanca: *Un tipo de cordón de soldadura depositado sin zigzagueo notable.*

stub: The short length of filler metal electrode, welding rod, or brazing rod that remains after its use for welding or brazing.
colilla: *El pedazito de electrodo con metal de aportación o varilla de soldadura, que queda después de haber sido usados para hacer la soldadura.*

substrate: Any material to which a thermal spray deposit is applied.
substrato: *Todo material al que se le ha aplicado un depósito con rocío térmico.*

surface preparation: The operation necessary to produce a desired or specified surface condition.
preparación de superficie: La operación necesaria para producir una condición de superficie deseada o especificada.

surfacing: The application by welding, brazing, or thermal spraying of a layer of material to a surface to obtain desired properties or dimensions, as opposed to making a joint.
alisamiento: La aplicación, por medio de soldaduras, o rocío térmico, de un estrato de material a una superficie para obtener ciertas dimensiones; en vez de soldar una juntura.

surfacing weld: A weld applied to a surface, as opposed to making a joint, to obtain desired properties or dimensions.
cordón de alisamiento: Un cordón aplicado a una superficie – en vez de para hacer una juntura – para obtener ciertas propiedades o dimensiones.

sweat soldering: A soldering process variation in which workpieces that have been pre-coated with solder are reheated and assembled into a joint without the use of additional solder (also called sweating).
sudar: Un proceso de soldadura en el que piezas que han sido previamente cubiertas con soldadura, son recalentadas y montadas como una juntura sin tener que usar soldadura adicional.

T

tack weld: A weld made to hold the parts of a weldment in proper alignment until the final welds are made.
soldadura provisional a puntos: Una soldadura que es hecha solamente para tener las partes de una soldadura en alineamiento hasta que la soldadura final sea hecha.

tensile strength: The resistance to breaking exhibited by a material when subjected to a pulling stress. Measured in lb/in^2 or kPa.
resistencia a la tensión: La resistencia de un material a romperse cuando fuerzas tensiles están actuando sobre él. Sus unidades son: lb/in^2 o KiloPascal.

thermal conductivity: The ability of a material to transmit heat.
conductividad de calor: La habilidad de un material a transmitir energía calórica.

thermal cutting (TC): A group of cutting processes that severs or removes metal by localized melting, burning, or vaporizing of the workpieces.
corte térmico: Un grupo de procesos para cortar que separa o remueve metal fundiendo el metal localmente, quemando, o evaporando las piezas.

thermal expansion: The expansion of materials caused by heat input.
expansión térmica: La expansión de elementos debida a la absorción de calor.

thermal stress relieving: A process of relieving stresses by uniform heating of a structure or a portion of a structure, followed by uniform cooling.
reducción de esfuerzos térmicos: El proceso para aliviar los esfuerzos en una estructura o parte de ella, calentándola uniformemente seguido de un enfriamiento uniforme.

tinning: A non-standard term for pre-coating.
"estañar": Término, no aprobado, que quiere decir revestimiento preliminar.

T-joint: A joint between two members located approximately at right angles to each other in the form of a T.
juntura en T: Una juntura en la que los miembros se colocan formando la figura de una T aproximadamente .

241

torch brazing (TB): A brazing process that uses heat from a fuel gas flame.
soldadura fuerte con soplete: *Un proceso de soldadura fuerte en la que el calor viene de la llama de un gas combustible.*

torch oscillation: Moving a torch in a back and forth motion.
oscillación del soplete: *Movimiento del soplete con una oscilación adelante-atrás.*

torch soldering (TS): A soldering process that uses heat from a fuel gas flame.
soldadura blanda con soplete: *Un proceso de soldadura blanda en la que el calor viene de la llama de un gas combustible.*

transferred arc: A plasma arc established between the electrode of the plasma arc torch and the workpiece.
arco transferido: *Un arco de plasma establecido entre el electrodo del arco de plasma y la pieza.*

transverse crack: A crack with its major axis oriented approximately perpendicular to the weld axis.
grieta transversal: *Grieta cuyo eje mayor es aproximadamente perpendicular al eje de la soldadura.*

travel angle: The angle less than 90° between the electrode axis and a line perpendicular to the weld axis, in a plane determined by the electrode axis and the weld axis. This angle can also be used to partially define the positions of welding guns, torches, rods, and beams.
ángulo de viaje: *El ángulo (menos de 90°) entre el eje del electrodo y una línea perpendicular al eje de la soldadura, en un plano determinado por las dos líneas. Este plano puede usarse para definir, parcialmente, la posición de pistolas de soldar, sopletes, varas y vigas.*

tungsten electrode: A non-filler metal electrode used in arc welding, arc cutting, and plasma spraying, made principally of tungsten.
electrodo de tungsteno: *Electrodo metálico, sin material de aportación, usado en solda-dura con arco, corte con arco, rociado de plasma, y hecho principalmente de tungsteno.*

U

U-groove weld: A type of groove weld.
soldadura de ranura en U: *Un tipo de soldaduras de ranura.*

under-bead crack: A crack in the heat-affected zone generally not extending to the surface of the base metal.
grieta debajo del cordón: *Una grieta en la zona afectada por el calor, y que general-mente no llega hasta la superficie del material de base.*

undercut: A groove melted into the weld face or root surface and extending below the adjacent surface of the base metal.
rebaja: *Una ranura fundida dentro de la cara del cordón o la superficie de la raíz, y que se extiende por debajo de la superficie adyacente del metal de base.*

underfill: A condition in which the weld face or root surface extends below the adjacent surface of the base metal.
bajo-relleno: *Una condición en la que la cara del cordón o de la superficie de la raíz se extiende por debajo de la superficie del metal de base.*

uphill: Welding with an upward progression.
hacia arriba: *Soldadura vertical que procede de abajo hacia arriba.*

V

vertical welding position: The welding position in which the weld axis, at the point of welding, is approximately vertical, and the weld face lies in an approximately vertical plane.

posición de soldadura vertical: La posición en la que el eje de soldadura, en este punto de la soldadura, es aproximadamente vertical, y la cara del cordón resta en un plano aproximadamente vertical.

vertical up: A nonstandard term for uphill welding.

verticalmente: Un término, no aprobado, equivalente a soldadura hacia arriba.

V-groove weld: A type of groove weld.

soldadura de ranura en V: Un tipo de soldadura de ranura.

volt: A unit of electrical force or potential.

voltio: Unidad de tensión eléctrica o diferencia de potencial entre dos puntos de un circuito eléctrico.

W

waster plate: A piece of metal used to initiate thermal cutting.

plato de ayuda: Pedazo de plancha que sirve para iniciar el corte térmico.

watt: A unit of electric power equal to voltage multiplied by amperage. One horsepower is equal to 746 watts.

vatio: Unidad de potencia eléctrica. Se obtiene multiplicando la tensión por la corriente. Un "caballo inglés" (hp) es igual a 746 vatios.

weave bead: A type of weld bead made with transverse oscillation.

cordón zigzag: Un tipo de soldadura en la cual se hace oscilar a la boquilla en dirección transversal a la dirección de progreso de la soldadura.

weld: A localized coalescence of metal or nonmetals produced either by heating the materials to the welding temperature, with or without the application of pressure, or by the application of pressure alone, with or without the use of filler material.

soldadura: La coalescencia local de metales o metaloides, el resultado de haber calentado los materiales a la temperatura de soldadura, y con o sin la aplicación de presión; o del haber aplicado presión solamente, y con o sin el uso de materiales de aportación.

weldability: The capacity of material to be welded under imposed fabrication conditions into a specific suitably designed structure and to perform satisfactorily in the intended service.

soldabilidad: La capacidad de un material para ser soldado bajo las condiciones de fabricación estipuladas, para crear una estructura adecuada y que pueda desempeñar el servicio para el que fué diseñado satisfactoriamente.

weld axis: A line through the length of the weld, perpendicular to and at the geometric center of its cross-section.

eje de soldadura: Una línea que percorre la longitud de la soldadura atravesando per-pendicularmente el centro geométrico de cada sección transversal de la soldadura.

weld bead: A weld resulting from a pass.

botón de soldadura: El cordón de soldadura que resulta después de haber hecho un pase por la juntura.

weld crack: A crack located in the weld metal or heat-affected zone.

grieta de soldadura: Una grieta en el metal de soldadura o en la zona afectada por el calor.

welder certification: Written verification that a welder has produced welds meeting a prescribed standard of welder performance.

243

certificado de soldador: *Verificación, por escrito, que atesta que la persona ahí nombrada ha hecho soldaduras que satisfacen los estándares prescritos de ejecución.*

weld face: The exposed surface of a weld on the side from which welding was done.
cara del cordón: *La superficie externa del cordón de soldadura del lado por el que se hizo la soldadura.*

weld groove: A channel in the surface of a workpiece or an opening between two joint members that provides space to contain a weld.
ranura para soldadura: *Un pequeño canal o muesca en una pieza, o la apertura entre dos miembros de una juntura a soldar, que provee el espacio para hacer el cordón.*

welding: A joining process that produces coalescence of materials by heating them to the welding temperature with or without the application of pressure, or by the application of pressure alone with or without the use of filler metal.
soldadura: *Un proceso de unión de partes que resulta en la coalescencia de materiales calentándolos a la temperatura de soldadura, con o sin la aplicación de presión; o aplicando presión solamente; y con o sin el uso de material de aportación.*

welding arc: A controlled electrical discharge between the electrode and the workpiece that is formed and sustained by the establishment of a gaseous, conductive medium called an arc plasma.
arco para soldar: *Una descarga eléctrica, regulada, entre el electrodo y la pieza a soldar. Es generada y mantenida creando un medio gaseoso, conductor, que se llama arco de plasma.*

welding electrode: A component of the welding circuit through which current is conducted and that terminates at the arc, molten conductive slag, or base metal.
electrodo para soldar: *Un componente del circuito eléctrico para soldar, por el cual pasa una corriente que termina en el arco, o la escoria fundida, o el metal de base.*

244

welding filler metal: The metal or alloy to be added in making a weld joint that alloys with the base metal to form weld metal in a fusion welded joint.
metal de aportación para soldar: *El metal o liga de metales que se aporta cuando se va a hacer la soldadura de una juntura. Este material se liga con el metal de base para for-mar un metal de soldadura en lo que se llama una juntura soldada por fusión.*

welding helmet: A device equipped with a filter plate designed to be worn on the head to protect eyes, face, and neck from arc radiation, radiated heat, spatter, or other harmful matter expelled during some welding and cutting processes.
casco de protección: *Una parte del vestuario del soldador que se pone en la cabeza para proteger los ojos y el cuello contra radiación del arco, calor, chispas, u otro material que pueda brotar durante una soldadura o un corte con arco. Está equipado con un filtro de vidrio para los ojos.*

welding leads: The workpiece lead (cables) and electrode lead (cables) of an arc welding circuit.
alambres de conexión: *El par de cables que conectan los terminales del suministrador de energía con la pieza a soldar y con el electrodo en un circuito para soldar con arco.*

welding positions: The relationship between the weld pool, joint, joint members, and welding heat source during welding.
posiciones para soldar: *La relación espacial que existe entre la balsa de soldadura, la juntura, los miembros de la juntura, y el aparato suministrador de calor, durante la soldadura.*

welding power source: An apparatus for supplying current and voltage suitable for welding.
suministrador de energía para soldar: *Aparato que produce la tensión y la corriente adecuadas para soldar.*

welding procedure: The detailed methods and practices involved in the production of a weldment.

procedimiento para soldar: *Los métodos y la práctica, en detalle, que se emplean en la producción de una soldadura.*

welding rectifier: A device, usually a semiconductor diode, in a welding power source for converting alternating current to direct current.
rectificador para soldaduras: *Un dispositivo – generalmente un diodo semiconductor – en un suministrador de energía para soldaduras, que convierte corriente alterna en corriente contínua.*

welding rod: A form of welding filler metal, normally packaged in straight lengths, that does not conduct the welding current.
varilla de soldar: *Una forma de empaque del metal de aportación, normalmente en varas de cierta longitud, y que no conduce la corriente de soldadura.*

welding schedule: A written statement, usually in tabular form, specifying values of parameters and welding sequence for performing a welding operation.
plan de soldaduras: *Una planilla que enumera los valores de los parámetros y la secuencia de las soldaduras requeridas para hacer un trabajo específico.*

welding sequence: The order of making welds in a weldment.
secuencia de soldaduras: *El orden en que se deben hacer las soldaduras en cada caso.*

welding transformer: A transformer used for supplying current for welding.
transformador para soldar: *Un transformador diseñado para suministrar la corriente necesaria para soldar.*

welding wire: A form of welding filler metal, normally packaged as coils or spools, that may or may not conduct electrical current depending upon the welding process with which it is used.
alambre para soldar: *Una forma de metal de aportación, normalmente empacado en abobinado o en carretes que pueden ser – o no – conductores, según el tipo de soldadura en la que se usa.*

245

weld interface: The interface between weld metal and base metal in a fusion weld, between base metals in a solid-state weld without filler metal, or between filler metal and base metal in a solid-state weld with filler metal.
áreas colindantes: *El espacio entre el metal de la soldadura y el metal de base, en una soldadura a fusión; o entre metales de base en soldaduras de estado sólido pero sin metal de aportación; o entre metal de aportación y el metal de base en soldaduras de estado sólido con metal de aportación.*

weldment: An assembly whose component parts are joined by welding.
soldadura: *Un ensamblaje mantenido por la soldadura de sus miembros*

weld metal: The portion of a fusion weld that has been completely melted during welding.
metal de soldadura: *En soldaduras a fusión, la porción que se ha fundido completamente durante la soldadura.*

weld metal area: The area of weld metal as measured on the cross-section of a weld.
área de metal de soldadura: *El área de la sección de la soldadura*

weld pass: A single progression of welding along a joint. The result of a pass is a weld bead or layer.
pase de soldadura: *Una sola progresión de soldadura a lo largo de una juntura. El resultado es un cordón, o una capa, de soldadura.*

weld pass sequence: The order in which the weld passes are made.
secuencia de pases: *El número y el orden en que se deben hacer los pases en una soldadura.*

weld penetration: A nonstandard term for joint penetration and root penetration.

penetración de la soldadura: Un término, no aprobado, para penetración de juntura, o de raíz.

weld pool: The localized volume of molten metal in a weld prior to its solidification as a weld metal.
balsa de soldadura: El volumen de metal fundido , resultado de una soldadura, antes de que se solidifique, cuando se llama metal de soldadura.

weld puddle: A nonstandard term for weld pool.
charco de soldadura: Término, no aprobado, para balsa de soldadura.

weld reinforcement: Weld metal in excess of the quantity required to fill a joint.
refuerzo de soldadura: La cantidad de metal que excede la necesaria para llenar la juntura.

weld root: The points, shown in a cross-section, at which the root surface intersects the base metal surfaces.
soldadura de raíz: Los puntos, vistos en una sección, en que la superficie de la raíz toca las superficies del metal de base.

weld tab: Additional material that extends beyond either end of the joint, on which the weld is started or terminated.
pestaña: Material adicional que aparece a los extremos de la juntura, donde la soldadura comenzó y terminó.

weld toe: The junction of the weld face and the base material.
orilla de la soldadura: La zona donde la cara del cordón toca el material de base.

wetting: The phenomenon whereby a liquid filler metal or flux spreads and adheres in a thin continuous layer on a solid base metal.
mojar: El fenómeno que se muestra en la adherencia de metal de aportación en forma líquida o el fundente al metal de base, en una capa delgada y contínua a todo lo largo de la soldadura.

wiped joint: A joint made with solder having a wide melting range and with the heat supplied by the molten solder poured onto the joint. The solder is manipulated with a handheld cloth or paddle to obtain the required size and contour.
juntura limpiada: Una juntura unida con soldadura que tiene una gama muy amplia de temperaturas de fusión, y con el calor proveniente de soldadura fundida que se vierte sobre la juntura. La cantidad de soldadura puede ser ajustada pasando un trapo o una espátula por las orillas hasta obtener el tamaño y la forma deseada.

wire feed speed: The rate at which wire is consumed in arc cutting, thermal spraying, or welding.
velocidad de alimentación del alambre: La rapidez con que se consume el alambre en soldadura con arco, corte térmico, y otras soldaduras.

work angle: The angle less than 90° between a line perpendicular to the major work-piece surface and a plane determined by the electrode axis and the weld axis. In a T-joint or a corner joint the line is perpendicular to the non-butting member. This angle can also be used to partially define the positions of guns, torches, rods, and beams.
ángulo de trabajo: El ángulo (menos de 90°) entre una línea perpendicular a la superficie mayor de la pieza y el plano determinado por el eje del electrodo y el eje del cordón. En una juntura a T o en una juntura de esquina, la línea es perpendicular al miembro que no es de tope. Este plano puede usarse para definir, parcialmente, la posición de pistolas de soldar, sopletes, varas y vigas.

work hardening: Also called cold working; the process of forming, bending, or hammering a metal well below the melting point to improve strength and hardness.

246

endurecimiento por trabajo: *También llamado endurecimiento por trabajo en frío; es el proceso de martillear el metal a una temperatura muy baja con respecto a su tempera-tura de fusión, con el propósito de aumentar su dureza y su resistencia.*

workpiece: The part that is welded, brazed, soldered, thermal cut, or thermal sprayed.
pieza de trabajo: *La pieza que está siendo soldada, cortada o rociada térmicamente.*

work-piece lead: The electrical conductor between the arc welding current source and work-piece connection.
cable de la pieza de trabajo: *El cable aislado que conecta la pieza al suministrador de corriente de arco.*

wrought iron: A material composed almost entirely of iron, with very little or no carbon.
hierro forjado: *Material compuesto casi exclusivamente de hierro, y si acaso, con algunas trazas de carbono.*

Y

yield strength: The load at which a material will begin to yield, or permanently deform. Also referred to as yield point.
resistencia al relajamiento: *También llamado punto de relajamiento, es la magnitud de carga en tensión en la cual el material deja de ser elástico y comienza a deformarse permanentemente.*

247

INDEX

250

CREDITS

Photographs are courtesy of West Coast Customs, unless otherwise noted.
Illustrations are by Pamela Tallman and Lawrence Smith, except for those listed below.

From American Welding Society Publications:
Figures 1-5, 1-8, 1-10, 1-15, 1-18, 1-19, 1-20, 1-24, 1-28, 1-31, 3-14, 3-20, 4-1, 4-8, 5-1, 5-14, 5-17, 5-18, 5-19, 5-20, 5-21, 5-23, 6-3, 6-4, 6-9, 6-10, 6-11, 6-12, 6-15, 6-16, 7-3, 7-4, 7-12, 7-15, 7-16, 7-17, 7-18, 7-21, 8-1, 8-4b, 8-5, 8-6, 8-7, 8-9, 9-28, 9-29, 9-30, 9-31, 9-33, 9-34

Tables: 3-1, 4-2, 4-3, 5-1, 5-2, 5-5, 5-7, 5-8, 6-1, 6-2, 6-3, 8-1

ESAB Welding and Cutting Products:
Figures 3-19, 7-22, 9-26

Lincoln Electric Company:
Figures 4-14, 4-15, 4-16, 4-17, 4-18 Tables 4-1

Thermadyne Industries, Inc.:
Figures 3-23